T0305712

Introduction to Nuclear Engineering

Introduction to Nuclear Engineering serves as an accompanying study guide for a complete, introductory single-semester course in nuclear engineering. It is structured for general class use, alongside fundamental nuclear physics and engineering textbooks, and it is equally suited for individual self-study.

The book begins with basic modern physics with atomic and nuclear models. It goes on to cover nuclear energetics, radioactivity and decays, and binary nuclear reactions and basic fusion. Exploring basic radiation interactions with matter, the book finishes by discussing nuclear reactor physics, nuclear fuel cycles, and radiation doses and hazard assessment. Each chapter highlights basic concepts, examples, problems with answers, and a final assessment.

The book is intended for first-year undergraduate and graduate engineering students taking Nuclear Engineering and Nuclear Energy courses.

Dr. Supathorn Phongikaroon is an Engineering Foundation Professor and Nuclear Engineering Program Director in the Department of Mechanical and Nuclear Engineering at Virginia Commonwealth University (VCU). He earned his PhD and BS degrees in chemical engineering and nuclear engineering from the University of Maryland, College Park in 2001 and 1997, respectively. Prior to joining the VCU in January 2014, he held academic and research positions at the University of Idaho in Idaho Falls, ID; Idaho National Laboratory in Idaho Falls, ID; and Naval Research Laboratory, Washington, DC. During his research career, Dr. Phongikaroon established chemical and electrochemical separation of used nuclear fuel through pyroprocessing technology and extended his expertise toward reactor physics and material detection and accountability for safeguarding applications. His effort led to a strong establishment of Radiochemistry and Laser Spectroscopy Laboratories at VCU.

His work has been published in over 50 papers in peer-reviewed journals and presented at over 100 international and national conferences and workshops. He has been able to maintain continuous diverse research support from international and national programs through the Department of Energy, national laboratories, and other universities. He has been Principal Investigator and Co-Principal Investigator for more than 20 external supported projects in total of over 3.17 million dollars of external awards since joining academia in 2007. He has taught more than 30 classes in nuclear, mechanical and chemical engineering-related topics for resident students and more than 10 classes for continuing (long distance—video conference) education over the last 9 years.

Introduction to Nuclear Engineering
A Study Guide

Supathorn Phongikaroon

CRC Press
Taylor & Francis Group
Boca Raton London New York

CRC Press is an imprint of the
Taylor & Francis Group, an **informa** business

Designed cover image: Supathorn Phongikaroon

First edition published 2024
by CRC Press
2385 Executive Center Drive, Suite 320, Boca Raton, FL 33431, USA.

and by CRC Press
4 Park Square, Milton Park, Abingdon, Oxon, OX14 4RN

CRC Press is an imprint of Taylor & Francis Group, LLC

Library of Congress Cataloging-in-Publication Data
Names: Phongikaroon, Supathorn, author.
Title: Introduction to nuclear engineering : a study guide / Supathorn Phongikaroon.
Description: First edition. | Boca Raton : CRC Press, 2024. | Includes bibliographical references and index.
Identifiers: LCCN 2023011881 (print) | LCCN 2023011882 (ebook) |
ISBN 9781032224404 (hbk) | ISBN 9781032224411 (pbk) |
ISBN 9781003272588 (ebk)
Subjects: LCSH: Nuclear engineering—Textbooks.
Classification: LCC TK9146 .P56 2024 (print) | LCC TK9146 (ebook) |
DDC 621.48—dc23/eng/20230727
LC record available at https://lccn.loc.gov/2023011881
LC ebook record available at https://lccn.loc.gov/2023011882

ISBN: 978-1-032-22440-4 (hbk)
ISBN: 978-1-032-22441-1 (pbk)
ISBN: 978-1-003-27258-8 (ebk)

DOI: 10.1201/9781003272588

Typeset in Times
by codeMantra

Access the Instructor Resources: routledge.com/9781032224404

Contents

Preface

Nuclear physics and engineering topics can be daunting among newcomers. Many reference textbooks have been written rigorously to provide detailed information, which result in becoming intimidating to young scientists and engineers. In addition, it is extremely hard to cover all chapters in one introductory semester to nuclear engineering. The main purpose of this textbook is to provide a basic guidance to a complete year's course in many fundamental nuclear physics and engineering textbooks that are available in the field. This textbook is structured for general class use and for individual self-study.

The textbook is the outcome of over 10 years' work undertaken in the development of programmed learning techniques starting from the University of Idaho, Idaho Falls Campus to Virginia Commonwealth University. To motivate individuals in nuclear topics, this textbook is subdivided into eleven chapters that can be explored in one semester (15-week course module). The first three chapters start off with (Chapter 1) Basic Units and the Atoms leading to (Chapter 2) Basic Modern Physics with (Chapter 3) Atomic and Nuclear Models. The next three chapters cover (Chapters 4 and 5) Nuclear Energetics and (Chapter 6) Radioactivity and Decays. The subsequent chapters explore on (Chapter 7) Binary Nuclear Reactions, basic (Chapter 8) Radiation Interactions with Matter, and (Chapter 9) Neutron Chain Reactions and Basic Nuclear Reactor Physics. The last two chapters provide important topics on how the first nine chapters being applied to (Chapter 10) nuclear power and various types of nuclear reactors with a basic nuclear fuel cycle and how to deal with (Chapter 11) radiation doses and hazards. Each chapter is followed by a frame-by-frame learning method which highlights basic concepts, examples, problems with answers, and the end-of-module assessment through exercises. The advantages of working at one's own rate, the intensity of the student involvement, and the immediate assessment of responses are well known to those already acquainted with module learning activities.

I would like to thank my family, especially my wife and my daughter for giving me motivation to complete this textbook. Thank you to all my students—especially to Peggy Cawley and John Smith for helping me compiling all references and figures. I am super thankful to Professor Jim Miller for proofreading, correcting my problem designs, and extra notes, especially providing the drawing for the cover. Very significant acknowledgment belongs to Kyra Lindholm and others at CRC Press for believing in my work and for supporting this book's production.

1 Basic Units and the Atom

OBJECTIVES

After studying this chapter, the reader should be able to:

1. Understand and use modern units including the conversion factors.
2. Identify and provide proper use of significant figures in scientific calculations.
3. Know concept in nuclear physics—the energy with the electron volt unit.
4. Define atomic mass unit and utilize it in the calculation.
5. Define atom and its components including nuclear nomenclature and transform between the atomic and macroscopic worlds.
6. Interpret nuclear dimensions and approximations.

1.1 MODERN UNITS

SI system of metric units, also known as "International System of Units," has seven base units (Table 1.1), derived units (Table 1.2), supplementary units (Table 1.3), and temporary units (Table 1.4).

Example 1.1

Is velocity with the formula "m/s^{-1}" a (a) base SI unit, (b) derived unit, or (c) temporary unit?

1.2 DERIVED UNIT

The Answer to Example 1.1 is "derived unit." What is it? It can be seen that the derived units are those that are composed by the physical quantity based on the base SI units. Of course, the temporary units are those that need a special conversion to the value in SI unit. For example, when we are on a cruise and the

TABLE 1.1

Seven Base Units

Physical Quantity	Unit Name	Symbol
Length	meter	m
Mass	kilogram	kg
Time	second	s
Electric current	ampere	A
Thermodynamic temperature	kelvin	K
Luminous intensity	candela	cd
Quantity of substance	mole	mol

DOI: 10.1201/9781003272588-1

TABLE 1.2

Derived Units – Combinations of the Base Units

Physical Quantity	Unit Name	Symbol	Formula
Force	newton	N	$kg\ m\ s^{-2}$
Energy	joule	J	N m
Power	watt	W	$J\ s^{-1}$
Electric potential difference	volt	V	$W\ A^{-1}$
Electric resistance	ohm	Ω	$V\ A^{-1}$
Radioactive decay	becquerel	Bq	s^{-1}
Pressure	pascal	Pa	$N\ m^{-2}$

TABLE 1.3

Supplementary Units

Physical Quantity	Unit Name	Symbol
Plane angle	radian	rad
Solid angle	steradian	sr

TABLE 1.4

Temporary Units Developed for Different Applications

Physical Quantity	Unit Name	Symbol	Value in SI Unit
Length	nautical mile		1852 m
Pressure	bar	bar	0.1 MPa
Pressure	standard atmosphere	atm	0.101325 MPa
Cross-sectional area	barn	b	$10^{-24}\ cm^2$
Radioactive activity	curie	Ci	$3.7 \times 10^{10}\ Bq$
Absorbed radiation dose	gray	Gy	$1\ J\ kg^{-1}$

announced speed is referred by "knot," this is indeed a temporary unit because 1 knot = $1852/3600\ m/s^{-1} = 0.5144\ m/s^{-1}$.

1.3 SI PREFIXES

The SI units and their symbols can be scaled by using the SI prefixes to indicate extremely small and large quantities. The most common prefixes are listed in Table 1.5.

1.4 CONVERSION FACTORS

Have you ever heard of inch (in), foot (ft), electron volt, gallon, etc.? These are widely used units outside the SI. Therefore, it is necessary to have a conversion factor to

TABLE 1.5

Common Prefixes

Factor	Prefix	Symbol	Factor	Prefix	Symbol
10^{15}	peta	P	10^{-1}	deci	d
10^{12}	tera	T	10^{-2}	centi	c
10^{9}	giga	G	10^{-3}	milli	m
10^{6}	mega	M	10^{-6}	micro	μ
10^{3}	kilo	k	10^{-9}	nano	n
10^{2}	hecto	H	10^{-12}	pico	P
10^{1}	deca	da	10^{-15}	femto	F

TABLE 1.6

Common Conversion Factors Used in Physics and Engineering

Property	Unit	SI Equivalent
Length	in	$0.0254\,m$
	ft	$0.3048\,m$
Area	ft^2	$0.09290394\,m^2$
	acre	$4046.873\,m^2$
Volume	gallon (U.S.)	$0.0037854\,m^3$
	ft^3	$0.028316\,m^3$
Pressure	lb_f/in^2 (psi)	$6894.757\,Pa$
	mm Hg (@0 °C)	$133.322\,Pa$
Energy	eV	$1.60219 \times 10^{-19}\,J$
	Btu	$1054.35\,J$
	kWh	$3.6 \times 10^6\,J$
Force	lb_f	$4.448\,N$

convert some non-SI units to their SI equivalent. Some conversion factors are listed in Table 1.6.

Readers can find more conversion factors in other engineering textbooks.

1.5 SIGNIFICANT FIGURES

The number of **significant figures** implies to the *number of digits reported for the value of a measured or calculated quantity, indicating the precision of the value.* There are three general rules:

Rule 1: All digits are significant except zeros at the beginning of the number and possibly terminal zeros (one or more zeros at the end of a number).

For example, 8.12 cm, 0.812 cm, and 0.0000812 cm all contain **THREE** significant figures. Now, you can do this one: 4.125 m has _____

1.6   FOUR SIGNIFICANT FIGURES

Here, we can see that all four digits are significant in 4.125 m, but what about zero ending in the written number, that is, 4.1000 cm for example.

Does it have (1) two, (2) three, (3) four, or (4) five significant figures? Take an educated guess!

1.7   FIVE SIGNIFICANT FIGURES

That is correct. The answer is indeed "5 significant figures." Why? This is because of
   Rule 2: Terminal zeros ending at the right of the decimal point are significant.

Example 1.2

8.00 cm, 8.10 cm, and 81.0 cm have **THREE** significant figures.
Finish it off with the final rule in the next section.

1.8   FINAL RULE OF SIGNIFICANT FIGURES

Rule 3: Terminal zeros in a number without an explicit decimal point may or may not be significant. If someone gives a measurement as 200 cm, it is impossible to know whether or not it is intended for one, two, or three significant figures. If the person writes 200 cm (please note the decimal point), then the zeros are significant. More generally, you can remove any uncertainty in such cases by expressing the measurement in scientific notation.

Scientific notation *is the representation of a number in the form* $A \times 10^n$, *where A is the number with a single nonzero digit to the left of the decimal point and n is an integer (or a whole number).*

Example 1.3

700 cm can be written to two significant figures by 7.0×10^2.
700 cm can be written to three significant figures by 7.00×10^2.

Now you should have no trouble with these exercises:

How many significant figures are there in each of the following measurements? (1) 140.0 kg, (2) 0.0458 g, (3) 0.99800 m, (4) 1008 s, (5) 9.10×1020 cm/s, and (6) 72.0000 A.
When you have finished, move on to the next lesson.
Here are the answers. Check with yours:

   a. Based on Rules 1 and 2, the answer is **FOUR**.
   b. Based on Rule 1, the answer is **THREE**.
   c. Based on Rule 2, the answer is **FIVE**.
   d. Based on Rule 1, the answer is **FOUR**.
   e. Based on Rule 3, the answer is **THREE**.
   f. Based on Rules 1 and 2, the answer is **SIX**.

All correct? Right. Now turn on to the next section to continue the chapter.

1.9 SPECIAL NUCLEAR UNITS

Physical quantities (energies and masses) are very small in SI units in nuclear science and engineering. Therefore, special units are almost always used. There are two popular units.

1st unit: **Electron volt**—a special energy unit to describe the energy being released or absorbed during a chemical reaction. Here, it is equivalent to the kinetic energy gained by an electron that is being accelerated through a potential difference ΔV of one volt.

So, $1\ V = \dots$

$$1\ W\,/\,A = \frac{1\left(J\,/\,s\right)}{\left(C\,/\,s\right)} = 1\,J\,/\,C.$$

So, the work done by the electric field is

$$-e\Delta V = \left(1.60217646 \times 10^{-19}C\right)\left(1\ J/C\right) = 1.60217646 \times 10^{-19}\,J = 1\ eV.$$

This is a very important concept in nuclear physics associating the energy with the **electron volt**. Thus, if the work is being done on the electron to accelerate through a potential of 1 V, then we can expressed it as

$$KE = \frac{1}{2}m_e v^2 = -e\Delta V = \mathbf{1\ eV}.$$

Based on this information, we can determine the velocity of the electron by noting that the mass of electron is $9.10938188 \times 10^{-31}$ kg. Thus, the velocity of the electron to the three significant figures at 1 eV is ....

$$\mathbf{5.93 \times 10^5\ m/s.}$$

We can show how to obtain the above answer in detail here:

$$v = \sqrt{\frac{2KE}{m_e}} = \sqrt{\frac{2\left(1.60217646 \times 10^{-19}\,J\right)}{9.10938188 \times 10^{-31}\,kg}} = \sqrt{\frac{2\left(1.60217646 \times 10^{-19}\ \frac{kg \times m^2}{s^2}\right)}{9.10938188 \times 10^{-31}\ kg}}$$

$$\approx 5.93 \times 10^5\ m/s.$$

Now we are ready to learn about the second popular nuclear unit in the next section.

1.10 ATOMIC MASS UNIT

It is another standard unit independent of the SI kilogram mass standard, which is 1/12 of the mass of a neutral ground-state atom of carbon-12 (^{12}C). Here, the mass of $N_a\,^{12}C$ atoms (Avogadro's number = 1 mole) is 0.012 kg. Thus,

$$1 \text{ amu} = \left(\frac{1}{12}\right)(0.012 \text{ kg/N}_a) = 1.6605387 \times 10^{-27} \text{ kg}.$$

Example 1.4

The radius of an iron nucleus is about 4×10^{-15} m, and its mass is 56 amu. Find the average density of the nuclear material.

Solution

$$\rho = \frac{m}{V} = \frac{m}{\frac{4}{3}\pi r^3} = \frac{(56 \text{ amu})(1.66 \times 10^{-27} \text{ kg/amu})}{\frac{4}{3}\pi(4 \times 10^{-15} \text{ m})^3} = 3.467 \times 10^{17} \text{ kg/m}^3.$$

So, how many more times dense than water is this?
    When you have worked this question, let *check your answer.*
    Here is the general solution to the previous question. First, the density of water is 1000 kg/m³. Therefore,

$$\frac{\rho}{\rho_{water}} = \frac{3.467 \times 10^{17}}{1000} = 3.467 \times 10^{14}.$$

We start to see that there are several physical constants that we have to know relating to nuclear science and technology because we have to compare things to the speed of light in vacuum, electron charge, atomic mass unit, etc. These values are critical in our existing physical world.

1.11   PHYSICAL CONSTANTS

From Table 1.7, we can practice the conversion of the unit. That is,

    i. 1 electron: 9.10938×10^{-31} kg = u
    ii. 1 proton: 1.67262×10^{-27} kg = MeV/c²
    iii. Boltzmann constant (k): 1.38065×10^{-23} JK^{-1} = eVK^{-1}

Here are the results in detail:

i. 9.10938×10^{-31} kg $\times \dfrac{1 \text{ u}}{1.66054 \times 10^{-27} \text{ kg}} = \mathbf{5.48579 \times 10^{-4} \text{ u}}$

ii. 1.67262×10^{-27} kg $\times \dfrac{931.494 \text{ MeV/c}^2}{1.66054 \times 10^{-27} \text{ kg}} = \mathbf{938.272 \text{ MeV/c}^2}$

iii. 1.38065×10^{-23} J/K $\times \dfrac{1 \text{ eV}}{1.60218 \times 10^{-19} \text{ J}} = \mathbf{8.61734 \times 10^{-5} \text{ eVK}^{-1}}$

With these physical properties, we can explore some fun facts about energy and temperature prior to learning about the atom and its basics.

1.12 FUN FACT

Relationships between energy and temperature can be established because energy units can be expressed through temperature of a fluid under the Maxwell–Boltzmann energy distribution concept. The energy/temperature connection is a natural one since we realize that a fluid possesses internal energy by virtue of random (thermal) motion of individual particles. The Boltzmann constant can be redefined by scaling the energy/temperature in atomic dimensions, that is,

$$k \times \text{Temperature (K)} = \text{Temperature (K)} \times 1.38 \times 10^{-23} \text{ J/K} \times \frac{1 \text{ eV}}{1.6 \times 10^{-19} \text{ J}}$$

$$\approx \text{Temperature (K)} \times \frac{1 \text{ eV}}{11,600 \text{ K}}.$$

Therefore, a fluid at 11,600 K would possess the equivalent thermal energy per particle of ... eV.

It is also possible to find the relationship between Kelvin and electron volt.

Example 1.5

So, what is the eV equivalent of room temperature at 293.15 K? The answer can be directly determined to be

$$293.15 \text{ K} \left(\frac{1 \text{ eV}}{11,600 \text{ K}} \right) = 0.0253 \text{ eV}.$$

The value of 0.0253 eV will be significant in our later discussion, especially dealing with the life of neutrons. However, before we can proceed to that, we have to learn briefly about "the atom."

1.13 THE ATOM

The most basic of nuclear science and technology is the concept that "all matter is composed of many small discrete units of mass called **atoms**." A simplistic view of an atom is a very small dense **nucleus**, composed of protons and neutrons (together called **nucleons**), that is surrounded by negatively charged electrons. In a neutrally charged atom, the number of electrons and protons are the same and atoms cannot be seen by naked eye. Although atoms cannot be seen with the naked eye, it is possible to observe them using tunneling electron microscopes to generate electrical signals.

1.14 OPTIONAL ON ORDINARY MATTER

From the general engineering perspective, we will briefly discuss about the fundamental components of ordinary matters. A few basics are as follows:

- Neutrons and protons are composed of other smaller particles known as "quarks" which are held together by the exchange of massless particles called "gluons."

TABLE 1.7

Physical Constants

Constant	Value (Up to Six Significant Figures)	Symbol
Speed of light (in vacuum)	$2.99792 \times 10^8 \, \text{m/s}^{-1}$	c
Electron charge	$1.60218 \times 10^{-19} \text{C}$	e
Atomic mass unit (amu)	$1.66054 \times 10^{-27} \text{kg}$ (931.494 MeV/c²)	u
Electron rest mass	$9.10938 \times 10^{-31} \text{kg}$ (0.510999 MeV/c²)	m_e
Proton rest mass	$1.67262 \times 10^{-27} \text{kg}$ (938.272 MeV/c²)	m_p
Neutron rest mass	$1.67493 \times 10^{-27} \text{kg}$ (939.565 MeV/c²)	m_n
Avogadro's constant	$6.02214 \times 10^{23} \text{mol}^{-1}$	N_a
Boltzmann constant	$1.38065 \times 10^{-23} \text{J K}^{-1}$	k
Ideal gas constant (STP)	$8.31447 \, \text{J mol}^{-1} \, \text{K}^{-1}$	R
Planck's constant	$6.62607 \times 10^{-34} \text{J s}$	h

- Gluons are the carriers of the nuclear force.
- Protons and neutrons can be referred to as nucleons or "baryons."
- Leptons are electrons, positrons, and muons.

For a basic understanding of quarks and leptons, the readers are referred to the website: https://www.youtube.com/watch?v=pdVybAwVqUs.

1.15 ATOMIC AND NUCLEAR NOMENCLATURE

Each atom can be uniquely identified by the number of neutrons N and protons Z within its nucleus. For a neutral charge atom, the number of electrons is equivalent to the number of protons Z, also known as **atomic number**. All atoms of the same element will have the same atomic number. For example, all carbon atoms have six protons in the nucleus.

However, atoms of the same element may have different number of neutrons N. Atoms of the same element with different N are known as **isotopes**. A particular isotope is identified as follows:

- A is the **mass number** representing the number of nucleons in the nucleus; that is, $A = Z + N$.
- X is the chemical symbol for each element shown in the periodic table.

It is not necessary to specify Z as X and A are sufficient to identify the isotope.

Example 1.6

In the case of ^{235}U and ^{238}U, the atomic number for both isotopes is **92**.

It is strongly recommended that all beginners should learn the complete list of symbols to master the nuclear nomenclature.

Example 1.7

Identify the number of protons, neutrons, and electrons for the following nuclear nomenclatures, assuming neutrally charged atoms.

$$^{12}_{3}\text{Li}$$

$$^{39}_{13}\text{Al}$$

Solution

a. Lithium has three protons ($Z = 3$) and the mass number of 12 ($A = 12$), resulting in nine neutrons ($A - Z = N = 9$). The number of electrons is **3**.

b. Aluminum has 13 protons ($Z = 13$) and the mass number of 39 ($A = 39$), and 26 neutrons ($A - Z = N = 26$). The number of electrons is **13**.

Now you should have no trouble identifying the element, the atomic number (Z), mass number (A), and number of neutrons (N) of the following: (a) $^{236}_{92}\text{U}$, (b) ^{56}Fe, and (c) ^{18}O.

When you have finished, move on to the next frame.

1.16 HERE ARE THE RESULTS

• $^{236}_{92}\text{U}$ is uranium, with $Z = 92$, $A = 236$, and $N = 236 - 92 = 144$.

Answers:

URANIUM	Z = 92	A = 236	N = 144

• ^{56}Fe is iron. Now, if you don't know the number of protons associated with this element, you need to look it up in the periodic table. Here, it has 26 protons ($Z = 26$), $A = 56$, and $N = 56 - 26 = 30$.

Answers:

IRON	Z = 26	A = 56	N = 30

• ^{18}O is oxygen (using the same approach), with $Z = 8$, $A = 18$, and $N = 18 - 8 = 10$.

Answers:

OXYGEN	Z = 8	A = 18	N = 10

Now move on to the next section for more definitions.

1.17   DEFINITIONS

- **Nuclides** are a particular atom or nucleus with a specific number of neutrons N and atomic number Z.
  - **Stable nuclides** are nuclides that maintain the same N and Z.
  - **Radionuclide** will change to another nuclide with a different Z or N by emitting one or more particles over time.
- **"Iso"** is a prefix meaning "the same."
  - **Isobars** are nuclides with the same mass number A but differing number of neutrons and protons. *Example*: $^{14}_{5}$B, $^{14}_{7}$N, $^{14}_{8}$O.
  - **Isotones** are nuclides with the same number of neutrons but different numbers of protons. *Example*: $^{14}_{6}$C, $^{15}_{7}$N, $^{16}_{8}$O.
  - **Isotopes** are the same element (same Z) with different numbers of neutrons. *Example*: $^{234}_{92}$U, $^{235}_{92}$U, $^{238}_{92}$U.
  - **Isomers** are the same nuclide, and same Z and N in which the nucleus is in different long-lived *excited* states. We will discuss this later.

There is a trick in remembering isob**a**rs, isoto**n**es, and isoto**p**es. Noticing the underline letter for term. The "a, n, and p" represent mass number A, neutron, and proton, respectively. This makes it very easy to remember isobar (constant mass number A), isotone (constant neutron N), and isotope (constant proton Z).

Now turn on to the next section to continue the definitions.

1.18   MORE DEFINITIONS

- **Atomic weight (AW)** is the ratio of the atom's mass to one-twelfth that of one neutral ^{12}C atom.
- **Molecular weight (MW)** is the total of all the atomic weights of the atoms in a molecule.
- **Isotopic abundance** γ is a fraction or percentage of the atoms in representing the stable isotopes in each element .
- **Elemental atomic weight** is the weighted average of the atomic weights of all naturally occurring isotopes of the element; it is calculated as follows:

$$AW = \sum_i \frac{\gamma_i(\%)}{100} AW_i.$$

- **Avogadro's number (N_a)** is defined as 6.022×10^{23} atom/mol, as defined by a fundamental principle relating microscopic to macroscopic quantities. It is equivalent to number of atoms in 12 grams of ^{12}C. Here, one gram atomic or molecular weight of any substance represents one mole of the substance and contains as many atoms or molecules as there are atoms in one mole of ^{12}C.

From this aspect of definitions, you now should be able to do the following example.

Example 1.8

Show that atomic weight of X = amu of X.

Solution

Here, atomic weight of $X = \dfrac{\text{Mass of X}}{\left(\dfrac{\text{Mass of }^{12}\text{C atoms}}{12}\right)}$.

The mass of a neutral ^{12}C atom is 12 amu or 12 u. This implies that atomic weight of X = amu of X.

Before working on different problems, let's explore on how to use Avogadro's number and the atomic weight to transport between the macroscopic and microscopic/atomic worlds.

1.19   TRANSFORMATION BETWEEN THE ATOMIC AND MACROSCOPIC WORLDS

See Table 1.8.

The bottom-right corner quantity, $\dfrac{\rho}{AW} N_a \left(\dfrac{\text{atom}}{\text{cm}^3}\right)$, is known as **atomic number density**, N, which will be very useful in nuclear engineering calculation (discussed later in future chapters).

Example 1.9

What is the molecular weight (g/mol) of H_2SO_4?

Solution

From the Appendix I.B,

$$AW(H)=1.007825 \text{ g/mol} \quad AW(S)=32.065 \text{ g/mol} \quad AW(O)=15.9994 \text{ g/mol}$$

$$\text{Therefore, MW} = 2 \times AW(H) + 1 \times AW(S) + 4 \times AW(O)$$

$$= 98.0793 \text{ g/mol} \sim 98.08 \text{ g/mol} (\text{four significant figures}).$$

TABLE 1.8

Relationship between Macroscopic and Microscopic Scales

Macroscopic	Transformation	Microscopic World
m (g)	$\div AW \rightarrow \dfrac{m}{AW}(\text{mol}) \rightarrow \times N_a$	$\dfrac{m}{AW} N_a (\text{atom})$
$\rho\left(\dfrac{\text{g}}{\text{cm}^3}\right)$	$\div AW \rightarrow \dfrac{\rho}{AW}\left(\dfrac{\text{mol}}{\text{cm}^3}\right) \rightarrow \times N_a$	$\dfrac{\rho}{AW} N_a \left(\dfrac{\text{atom}}{\text{cm}^3}\right)$

We can further explore this problem by asking the following question:
What is the mass in kg of a molecule of H_2SO_4?
Well, think about it and once you are done with your calculation, check whether or not you are on the right track.
The solution is **1.6284×10^{-25} kg/molecule.**
How?
Here is the solution to the above answer.
We know that MW of H_2SO_4 is 98.0793 g/mol = 0.098079 kg/mol.
So, mass of one molecule = 0.098079 kg/mol × (1 mole/6.022 × 1023 molecules) = **1.6284×10^{-25} kg/molecule.**
Now, let's work on another example.

Example 1.10

What is the molecular weight (g/mol) of UO_2?
When you have finished, check your answer.
The answer is **270.028 g/mol.**
Here is the answer to the above question in detail.
From the Appendix, AW(U) = 238.0289 g/mol and AW(O) = 15.9994 g/mol. Therefore, MW = 1 × AW(U)+2 × AW(O) = **270.0277 g/mol** or **270.028 g/mol** (six significant figures).
Let's practice some more.

Example 1.11

5 g of boron = _____ atoms of ^{10}B

Solution

The atomic weight of boron is 10.811 g/mol as listed in Appendix I.B.
Therefore, 5 g of boron will contain

$$5\ g \times \frac{1\ mol}{10.811\ g/mol} \times \frac{6.022 \times 10^{23}\ atoms}{1\ mol} = 2.785 \times 10^{23}\ atoms.$$

Since the isotopic abundance of ^{10}B is 19.9% (Appendix I.B),

$$(2.785\ 10^{23}) \times (0.199) = \mathbf{5.542 \times 10^{22}}\ \textbf{atoms of } \mathbf{^{10}B}.$$

Example 1.12

The density of dry air at STP is typically 0.0012 g/cm³. Oxygen content in air is ~23% by weight. What is the atom density (atom/cm³) of ^{17}O?

Solution

The atom density of oxygen is

$$N_{oxygen} = \frac{\rho_{oxygen}N_a}{AW_{oxygen}} = \frac{w_{oxygen}\rho_{air}N_a}{AW_{oxygen}} = \frac{(0.23)(0.0012)(6.022 \times 10^{23})}{15.9994} = 1.0 \times 10^{19}\ atoms/cm^3.$$

Here, the isotopic abundance of ^{17}O in elemental oxygen is 0.038% of all oxygen atoms. Therefore, the atom density of ^{17}O is

$$\gamma_{17_O} \times N_{oxygen} = \frac{0.038}{100} \times 1.0 \times 10^{19} \text{ atoms/cm}^3 = 3.8 \times 10^{15} \text{ atoms/cm}^3.$$

1.20   MASS OF AN ATOM

We can see at this point that Avogadro's number plays a significant role in converting the mass in grams to atoms. In general, the mass of an individual atom can be expressed as

$$M(g/atom) = \frac{AW}{N_a} \approx \frac{A}{N_a}.$$

That is, the approximation of the atomic weight (AW) by the mass number (A) of that atom is quite acceptable for all but the most precise calculations. This approximation will be seen throughout this book.

Example 1.13

Estimate the mass of ^{16}O using both the isotopic atomic weight and the mass number.

Solution

The mass of ^{16}O is found to be 15.9949 g/mol; this is the elemental atomic weight (AW). Thus,

$$M(^{16}O) = \frac{15.9949 \text{ g/mol}}{6.022 \times 10^{23} \text{ atom/mol}} = 2.65607 \times 10^{-23} \text{ g/atom.}$$

This is rather a precise approach. Now, we can use the mass number A = 16 for the calculation; this becomes

$$M(^{16}O) = \frac{16 \text{ g/mol}}{6.022 \times 10^{23} \text{ atom/mol}} = 2.65692 \times 10^{-23} \text{ g/atom.}$$

To be correct, the calculation with the mass number A should have only two significant figures. But for the sake of comparison, we will keep six significant figures to show that the approximation by the mass number A = 16 leads to a very small and often negligible error (~0.0320%).
   Now you should have no trouble with these problems.

Example 1.14

(a) Nitrogen has two stable isotopes, ^{14}N with isotopic abundance of 99.632% and ^{15}N with isotopic abundance of 0.368%. Calculate the average atomic weight of nitrogen (u); (b) then, estimate the mass of ^{14}N (g/atom) using the average atomic weight of nitrogen; and (c) what is the ^{18}O atom density in water with a molecular density of 1 g/cm³?
   Here are the results in detail.

Solution:

a. Atomic masses of ^{14}N and ^{15}N are 14.003074 u and 15.000108 u, respectively. Thus, 0.99632(14.003074 u)+0.00368(15.000108) = **14.00674 u**. This agrees well with the value listed in the periodic table for nitrogen, 14.00674 u. This is equivalent to 14.00674 g/mol.

b. We can see from (a) that ^{14}N is the naturally occurring isotope, and thus, ^{14}N mass is 14.003074 u or 14.003074 g/mol. Thus,

$$M\left(^{14}N\right) = \frac{14.003074 \text{ g/mol}}{6.022 \times 10^{23} \text{ atom/mol}} = \mathbf{2.3253 \times 10^{-23} \text{ g/atom}}.$$

c. First, we have to determine the molecular weight of water, which is calculated as follows:

$$MW\left(H_2O\right) = 2 \times AW\left(H\right) + 1 \times AW\left(O\right) \approx 2 \times 1 + 16 = 18 \text{ g/mol}.$$

Thus, the molecular density of H_2O is calculated as

$$N(H_2O) = \frac{\rho_{H_2O} N_a}{MW_{H_2O}} = \frac{\left(1 \frac{g}{cm^3}\right)\left(6.022 \times 10^{23} \frac{\text{molecules}}{\text{mol}}\right)}{18 \frac{g}{\text{mol}}} = 3.35 \times 10^{22} \frac{\text{molecules}}{cm^3}.$$

Now, the oxygen density would be N(O) = 1 mole of $N(H_2O)$ = 3.35 × 10²² atoms/cm³. The isotopic abundance of ^{18}O is 0.205%; therefore,

$$N(^{18}O) = \left(\frac{0.205}{100}\right)\left(3.35 \times 10^{22} \frac{\text{atoms}}{cm^3}\right) = \mathbf{6.87 \times 10^{19} \frac{\text{atoms}}{cm^3}}.$$

1.21 NUCLEAR DIMENSIONS AND APPROXIMATIONS

If we assume that the volume of neutron and that of a proton are equal ($V_{neutron} = V_{proton}$), then the volume of a nucleus should be proportional to the mass number A. In particular, an effective spherical nuclear radius is often described by

$$R = R_0 A^{\frac{1}{3}}, \text{ with } R_0 \approx 1.25 \times 10^{-13} \text{ cm}.$$

The associated volume is

$$V_{nucleus} = \frac{4}{3}\pi R^3 \approx \left(7.25 \times 10^{-39}\right) \times A \text{ cm}^3.$$

We can see here that the nucleus volume is directly proportional to A. The mass density of a nucleus is

$$\rho_{nucleus} = \frac{m_{nucleus}}{V_{nucleus}} = \frac{\frac{A}{N_a}}{\left(\frac{4}{3}\right)\pi R^3} = 2.4 \times 10^{14} \text{ g/cm}^3.$$

Thus, the nucleus density is equal to the density of the earth if it were compressed to a ball roughly 200 m in radius.

Example 1.15

Based on the NASA information (https://nssdc.gsfc.nasa.gov/planetary/factsheet/earthfact.html), the earth equatorial radius is about 6378 km with its mass of 5.97×10^{24} kg. Determine the earth radius if it has the same mass density as matter in a nucleus.

Solution

We know that $\rho_{nucleus} = \dfrac{m_{nucleus}}{V_{nucleus}} = \dfrac{\dfrac{A}{N_a}}{\left(\dfrac{4}{3}\right)\pi R^3} = 2.4 \times 10^{14}$ g/cm^3 and the fact that

mass = density × volume, where the volume is $V = (4/3)\pi R^3$, where R is the earth radius in this case.

$$R = \left(\frac{3 \times mass}{4\pi\rho_{nucleus}}\right)^{\frac{1}{3}} = \left(\frac{3 \times 5.97 \times 10^{27}\ g}{4\pi \times 2.4 \times 10^{14}\ \dfrac{g}{cm^3}}\right)^{\frac{1}{3}} = \left(5.938 \times 10^{12}\ cm^3\right)^{\frac{1}{3}} = \mathbf{18108.88\ cm}.$$

This is **181.88 m** which is roughly 200 m in radius. That is, the radius of the earth is being compressed into a sphere with the density of a nucleus.

Important terms—in the order of appearance

SI units	Atomic number	Isomers
Significant figures	Isotopes	Atomic weight
Scientific notation	Mass number	Molecular weight
Electron volt	Nuclides	Isotopic abundance
Atomic mass unit	Stable nuclides	Elemental atomic weight
Atom	Radionuclide	Avogadro's number
Nucleus	Isobars	Atomic number density
Nucleon	Isotones	

BIBLIOGRAPHY

Bush, H.D., *Atomic and Nuclear Physics,* Prentice Hall, Englewood Cliffs, NJ, 1962.

Ebbing, D.D., *General Chemistry 5th Ed.,* Houghton Mifflin Company, Boston, MA, 1996.

Foster, A.R. and R.L. Wright Jr., *Basic Nuclear Engineering 4th Ed.,* Allyn and Bacon, Newton, MA, 1983.

Mayo, R.M., *Nuclear Concepts for Engineers,* American Nuclear Society, La Grange Park, IL, 1998.

Shultis, J.K. and R.E. Faw, *Fundamentals of Nuclear Science and Engineering 2nd Ed.,* CRC Press Taylor & Francis Group, Boca Raton, FL, 2008.

Shultis, J.K. and R.E. Faw, *Radiation Shielding,* American Nuclear Society, La Grange Park, IL, 2000.

FURTHER EXERCISES

A. True or False: If the statement is false, give a counterexample or explain the correction. If the statement is true, explain why it is true.

1. The atomic nucleus contains protons, neutrons, and electrons.
2. ^{16}O is observed to have the highest natural abundance on any oxygen isotope.
3. Becquerel is the unit name describing radioactive decay rate.
4. Luminous intensity is one of the physical quantities in the base SI units.
5. Isobars are nuclides with the same number of neutrons.
6. Isotopes are nuclides with the same number of protons.
7. $1000\,m$ can be written to 3 significant figures as $1000 \times 10^3\,m$.
8. The mass number of ^{232}Np is 232 u.
9. The radius of a carbon nucleus and its mass are $3 \times 10^{-15}\,m$ and 12 u, respectively. Thus, it is about 1.8×10^{12} denser than water with a density of $1000\,kg/m^3$.
10. Avogadro's constant represents the number of atoms in 1 g of ^{12}C.

B. Problems:

1. Explain the difference and similarity between these two units—hertz and curie.
2. What is the number of neutrons in ^{240}Pu? ^{208}Pb?
3. What is the mass in kg of a molecule of uranium oxide, U_3O_8?
4. How many atoms of ^{235}U are there in 1 g of natural uranium?
5. The density of natural uranium oxide (UO_2) is about 10.50 g/cm^3. Natural uranium is composed of three isotopes ^{234}U, ^{235}U, and ^{238}U.
   a. Compute the atomic weight of natural uranium (g/mol).
   b. What is the molecular weight of UO_2 fuel (g/mol)?
   c. What is the atomic density of ^{234}U in the UO_2 fuel (atoms/cm^3)?
6. How many atoms of deuterium are there in 2 kg of water assuming the natural abundance of deuterium in atom percent is 0.0115%?
7. Mars has a radius of about 3396 km and a mass of 6.417×10^{23} kg. Determine the radius of Mars (m) when it has the same mass density as matter in a nucleus.
8. A uranium sample is enriched to 3.2 atom percent in ^{235}U. Determine the enrichment of ^{235}U in weight-percent. Consider ^{234}U to be very small that can be ignored for the calculation.

2 Basic Modern Physics

OBJECTIVES

After studying this chapter, the reader should be able to:

1. Gain basic understanding of Einstein's theory of special relativity and learn about the difference between relativistic and nonrelativistic cases.
2. Understand the concept of radiation as waves and particles and quantum nature of light through photoelectric effect and Compton scattering.
3. Know about photons and how to use Einstein's photoelectric equation.
4. Learn about X-rays and behaviors under conservations of momentum and energy.
5. Define Compton wavelength and utilize it in the calculation.
6. Understand basic quantum theory dealing with the de Broglie wavelength, wave equation, wave–particle duality, and the uncertainty principle.

2.1 INTRODUCTION

There are three pillars in modern physics. These are (1) Einstein's Theory of Special Relativity, (2) Wave–Particle Duality, and (3) Quantum Mechanics. We will explore and give basic ideas and concepts for these important themes.

2.2 SPECIAL THEORY OF RELATIVITY

Newton's second law states that the rate of change in momentum $\mathbf{p}$ of the body is equivalent to the force $\mathbf{F}$ applying to it; that is,

$$\mathbf{F} = \frac{d\mathbf{p}}{dt} = \frac{d(m\mathbf{v})}{dt}.$$

It is important to see that under a constant mass assumption, the acting force $\mathbf{F}$ is simply a product between the mass m and the acceleration of the body $\mathbf{a}$, which is $\mathbf{F} = m\mathbf{a}$, where a is mainly the rate of change in velocity v of that body ($\mathbf{a} = d\mathbf{v}/dt$). It should be noted here that F, p, a, and v are in vector format; this means that they have directions.

In 1905, Einstein found an error in the above classical mechanics, $\mathbf{F} = m\mathbf{a}$. He showed that mass is in fact dependent upon velocity and can be expressed according to

$$m = \frac{m_0}{\sqrt{1 - \dfrac{v^2}{c^2}}},$$

where m_0 is the rest mass of the body and c is the speed of light under vacuum (3.0×10^8 m/s).

DOI: 10.1201/9781003272588-2

Example 2.1

Determine the fractional increase in mass of the ATLAS V ROCKET going at a speed of 16.26 km/s.

Solution

We can determine the fractional increase by observing the change in mass; that is,

$$\frac{m-m_0}{m_0} = \frac{\frac{m_0}{\sqrt{1-\frac{v^2}{c^2}}} - m_0}{m_0} = \frac{m_0\left(\frac{1}{\sqrt{1-\frac{v^2}{c^2}}} - 1\right)}{m_0} = \frac{1}{\sqrt{1-\frac{v^2}{c^2}}} - 1.$$

What we can see is that the change is dependent on the velocity ratio. So, we have to determine and analyze what we see. We must note that

$$\left(\frac{v}{c}\right)^2 = \left(\frac{16.26\times10^3 \text{ m}}{2.998\times10^8 \text{ m}}\right)^2 = 2.941\times10^{-9}.$$

This is quite a small number and to most calculators out there, it may return a value of 0 for the fractional mass increase. Therefore, we must apply a trick for evaluating relativistic expression like this (or for small values of v^2/c^2). We must recall the expression of $(1 + x)^n$ using a Taylor series expansion:

$$(1+x)^n = 1+nx+\frac{n(n+1)}{2!}x^2+\cdots \approx 1+nx \text{ for } |x| \ll 1.$$

Therefore, with $x = v^2/c^2$ and $n = -1/2$, we would have

$$\left(1-\frac{v^2}{c^2}\right)^{-1/2} \approx 1+\frac{1}{2}\times\frac{v^2}{c^2}.$$

Now, we can see that

$$\frac{m-m_0}{m_0} = \frac{1}{\sqrt{1-\frac{v^2}{c^2}}} - 1 = \left[1+\frac{1}{2}\times\frac{v^2}{c^2}\right] - 1 = \frac{1}{2}\times\frac{v^2}{c^2} = 1.4705\times10^{-9}.$$

What you can see here is that relativistic effects can safely be ignored in practical engineering problems. However, at the atomic and nuclear level where the velocity of the atoms can be as fast as the speed of light, the relativistic effect can be very important. Now let's move to the next frame.

2.3 E = mc² AND ITS PROOF

The important result of the special theory of relativity from Einstein's work is the derivation of $E = mc^2$ showing the equivalence of mass and energy E. Here, mass can be envisioned as a collection of particles and energy can be envisioned as waves. When dealing with engineering problems in the macroscopic world, this mass difference is rather insignificant. Yet, at the atomic and nuclear levels, this change is extremely vital.

To prove this, we must understand that the mass of the moving body would increase toward infinity as its traveling velocity increases toward the speed of light. Unless the velocity is greater than 10% of the speed of light, the correction is not important to engineering calculation (as illustrated in frame #3). Here, the kinetic energy, KE, is determined by subtracting the rest mass energy from the total energy:

$$\text{KE} = (m - m_0)c^2 = m_0 c^2 \left[\frac{1}{\sqrt{1 - \left(\frac{v^2}{c^2}\right)}} - 1 \right].$$

When $v \ll c$, the first two terms of the binomial expansion, that is $\left(1 + \frac{1}{2} \times \frac{v^2}{c^2} + \cdots\right)$, may be used for $1/\sqrt{1 - (v^2/c^2)}$ to yield $1/\sqrt{1 - (v^2/c^2)} = 1 + \frac{1}{2} \times \frac{v^2}{c^2} + \cdots$.

Substituting this into the original expression of KE gives the familiar expression for nonrelativistic kinetic energy, KE_{non}: $\text{KE}_{\text{non}} = m_0 c^2 \left(1 + \frac{1}{2} \times \frac{v^2}{c^2} - 1\right) = m_0 \frac{v^2}{2}$.

Example 2.2

Find the speed that a proton must be given if its mass is to be twice its rest mass of 1.67×10^{-27} kg. What energy (MeV) must be given the proton to achieve this speed?

Solution

We can use Einstein's Special Theory of Relativity again here by using $m = m_0 / \sqrt{1 - \frac{v^2}{c^2}}$. So, for $m = 2m_0$, this means that

$$2m_0 = \frac{m_0}{\sqrt{1 - \frac{v^2}{c^2}}} \Rightarrow 2 = \frac{1}{\sqrt{1 - \frac{v^2}{c^2}}} \Rightarrow 4\left(1 - \frac{v^2}{c^2}\right) = 1 \frac{v^2}{c^2} = \frac{3}{4} \Rightarrow v = 0.866\,c.$$

We can see that the proton must be traveling at least 86.6% of the speed of light. Now, in order to determine the energy, we need to see the change in the kinetic energy, KE, which is

$$\Delta KE = (m - m_0)c^2 = (2m_0 - m_0)c^2 = (1.67 \times 10^{-27} \text{ kg})(3 \times 10^8 \text{ m/s})^2$$

$$= 1.50 \times 10^{-10} \text{ J} = 9.38 \times 10^8 \text{ eV} = \textbf{938 MeV}.$$

So, it will take 938 MeV to achieve this speed. Readers often see the use of "MeV" in atomic/nuclear energy calculation.

2.4 RELATIONSHIP BETWEEN KINETIC ENERGY AND MOMENTUM

It is important to mention about the relationship between kinetic energy and momentum. That is, in classical physics, KE = 1/2 (mv²) = p²/(2m). Therefore, for relativistic particles, we can rearrange and manipulate $m = m_0 / \sqrt{1 - \dfrac{v^2}{c^2}}$ to get $m^2 \dfrac{c^2 - v^2}{c^2} = m_0^2$. Now, we can rearrange and show that

$$p^2 = \frac{1}{c^2}\left[(mc^2)^2 - (m_0 c^2)^2\right].$$

From this result, we can now have the expression for relativistic particles:

$$p^2 = \frac{1}{c}\sqrt{KE^2 + 2KE \times m_0 c^2}.$$

Readers will find this equation useful in calculating for KE in later chapters.

2.5 RADIATION AS WAVES AND PARTICLES

The energy in electromagnetic radiation is not continuous but comes in quanta with energies given by $E = h\nu = \dfrac{hc}{\lambda}$, where ν is the frequency, λ is the wavelength, and h is Planck's constant (h = 6.626 × 10⁻³⁴ J s = 4.136 × 10⁻¹⁵ eV s). The quantity hc occurs often in calculations and has the value of hc = 1240 eV nm (Note: the unit is electron volt-nanometers).

The quantum nature of light is exhibited in the

- **Photoelectric effect**, in which a photon is absorbed by an atom with the emission of an electron, and in
- **Compton scattering**, in which a photon collides with a free electron and emerges with reduced energy and therefore a greater wavelength.

We will explore about these types of photon interactions in the following sections.

2.6 PHOTOELECTRIC EFFECT

This was discovered by Hertz in 1887 and was studied by Lenard in 1900. When light is incident on the clean metal surface of the cathode, electrons are emitted (see Figure 2.1). If some of these electrons strike the anode, there is a current in

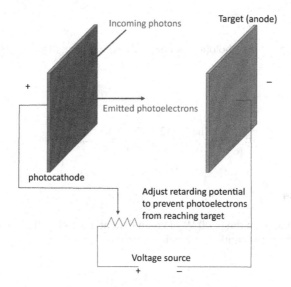

FIGURE 2.1 Photoelectric effect illustration.

the external circuit. The number of emitted electrons that reach the anode can be increased or decreased by making the anode positive or negative with respect to the cathode.

If we let V be the difference in potential between the cathode and the anode, then when V is positive, the electrons are attracted to the anode. At sufficient large value of V, all the emitted electrons reach the anode, and the current is at its maximum value. It should be noted that a further increase in V will not affect the current. Lenard observed that the maximum current is proportional to the light intensity. When V is negative, the electrons are repelled from the anode. Only electrons with initial kinetic energies that are greater than |eV| can then reach the anode.

In 1905, Einstein demonstrated that this experimental result can be explained if light energy is not distributed continuously in space but rather is quantized in small bundles called **photons**. The energy of each photon is $h\nu$. An electron emitted from a metal surface exposed to light receives its energy from a single photon. When the intensity of light of a given frequency is increased, more photons fall on the surface in unit time, but the energy absorbed by each electron is unchanged.

If Ξ is the energy is the energy necessary to remove an electron from a metal surface, the maximum kinetic energy of the electrons emitted will be $E_{max} = eV_0 = h\Xi - \Xi$, where Ξ is called the **work function**, which is a characteristic of the particular metal. This equation is also known as **Einstein's photoelectric equation**. Definitely, experimental verification of Einstein's theory was quite difficult. However, careful experiments by R.C. Millikan reported first in 1914 and later in more detail in 1916 showed that Einstein's equation was correct and that measurements of h agreed with the value found by Planck.

From this first lesson, we can see that a photon can be considered as an emission and absorption of electromagnetic energy with the value of $h\nu$.

Example 2.3

Find the energy of the photons in a beam whose wavelength is 526 nm.

Solution

Here,

$$E = hv = \frac{(hc)}{\lambda} = \frac{(1240 \text{ eV} \times \text{nm})}{526 \text{ nm}} = 2.36 \text{ eV} .$$

Example 2.4

Imagine a source emitting 200 W of green light at a wavelength of 500 nm. How many photons per second are emerging from the source?

Solution

Here, the power multiplied by the given time interval is the emitted energy, that is, (200 W)(1 s) = 200 J (use the conversion factor and units from Chapter 1). Denoting the photon flux by G, we have

$$G = \frac{200}{hv} = \frac{200\lambda}{hc} = \frac{(200 \text{ J})(500 \times 10^{-9} \text{ m})}{(6.6 \times 10^{-34} \text{ Js})(3 \times 10^8 \text{ m/s})} = 50 \times 10^{19} \text{ photons/s}.$$

Example 2.5

What is the momentum–energy relation for photons?

Solution

From Section 2.4, we know that $p^2 = \frac{1}{c^2}\left[(mc^2)^2 - (m_0 c^2)^2\right]$, which is $(pc)^2 = E^2 - (m_0 c^2)^2$. But a photon has zero rest mass; thus, **E = pc**.

Example 2.6

Find the largest momentum (kg m s^{-1}) that one can expect for a microwave photon (note: microwave frequencies can go up to 3 × 10^{11} Hz).

Solution

Since p=h/λ=(hv)/c,

$$p = \frac{(6.63 \times 10^{-34} \text{ Js})(3 \times 10^{11} \text{ s}^{-1})}{3 \times 10^8 \text{ m/s}} = 6.63 \times 10^{-31} \text{ kg ms}^{-1}.$$

Example 2.7

Calculate the work function of sodium metal if the photoelectric threshold wavelength is 500 nm.

Solution

Here, the work function Ξ is

$$h\nu_{min} = \frac{hc}{\lambda_{max}} = \frac{1240 \text{ eV nm}}{500 \text{ nm}} = \mathbf{2.48} \text{ } \mathbf{eV}.$$

Example 2.8

A photon of energy 4.0 eV imparts all its energy to an electron that leaves a metal surface with 1.0 eV of kinetic energy. Calculate the work function of the metal.

Solution

This is a straight forward approach; that is, $\Xi = h\nu - E_{max} = 4.0 - 1.0 = \mathbf{3.0} \text{ } \mathbf{eV}$.
Now you should have no trouble with these problems. Think carefully!

PROBLEM 2.1

A sensor with an opening diameter of 20 mm is exposed for 1 s to a 200 W lamp 20 m away. Assume that all the energy of the lamp is given off as light and the wavelength of the light is 500 nm. (a) Calculate the photon energy of that light (J). (b) Determine the number of photons emitted per second based on the lamp power. (c) Determine the number of photons that enter the sensor in 1 s.

PROBLEM 2.2

Find the electromagnetic radiation wavelength such that a photon in the beam is to have the same momentum as an electron moving with the speed of 2.5×10^5 m/s.

PROBLEM 2.3

Will photoelectrons be emitted by a copper surface with a work function of 4.4 eV when illuminated by visible light? Note: The wavelength range of the visible light is 400–700 nm.

When you have finished, move on to see the answer and how to solve this problem.
Answers
 Problem 2.1: (a) 3.98×10^{-19} J. (b) 5.03×10^{20} photons/s. (c) 3.1×10^{13} photons
 Problem 2.2: 2.91 nm
 Problem 2.3: Visible light cannot eject photoelectrons from this copper

SOLUTION TO PROBLEM 2.1

a. The energy of a photon of the light is $E = \dfrac{hc}{\lambda} = \dfrac{(6.63 \times 10^{-34} \text{ Js})(3 \times 10^{8} \text{ m/s})}{500 \times 10^{-9} \text{ m}} =$
   3.98×10^{-19} **J.**

b. Since the lamp uses 200 W of power, the number of photons emitted per
   second is $n = \dfrac{200 \text{ J/s}}{3.98 \times 10^{-19} \text{ J}} = \mathbf{5.027 \times 10^{20}}$ **photons/s.**

c. This last question is a bit tricky in thinking. We have to think that
   the radiation is spherically symmetrical and the number of photons entering the sensor per second is "n" multiplied by the ratio of the
   aperture area to the surface area of a sphere of radius 20 m. That is,
   $\left(5.027 \times 10^{20} \text{ photons/s}\right) \dfrac{\pi(0.010 \text{ m})^2}{4\pi(20 \text{ m})^2} = 3.142 \times 10^{13}$ photons/s.

Therefore, the number of photons that enter sensor in 1 s is $(1 \text{ s})(3.142 \times 10^{13}$
photons/s$) = \mathbf{3.142 \times 10^{13}}$ **photons.**

SOLUTION TO PROBLEM 2.2

It is required that $(mv)_{\text{electron}} = (h/\lambda)_{\text{photon}}$. Thus,

$$\lambda = \frac{h}{m_e v} = \frac{6.63 \times 10^{-34} \text{ Js}}{\left(9.1 \times 10^{-31} \text{ kg}\right)\left(2.5 \times 10^{5} \text{ m/s}\right)} = 2.91 \times 10^{-9} \text{ m} = \mathbf{2.91 \ nm}.$$

Note that mass of the electron can be looked up from Chapter 1. It should be noted
that this wavelength is in the X-ray region.

SOLUTION TO PROBLEM 2.3

We know that the photon energy just equals the energy required to tear the electron loose from the metal at threshold, namely, the work function Ξ. Thus, threshold
wavelength is threshold $\lambda = \dfrac{hc}{\Xi} = \dfrac{1240 \text{ eV nm}}{4.4 \text{ eV}} = 281.81 \text{ nm} \sim 282 \text{ nm}$. Since the
visible light is ranging between 400 nm and 700 nm, it **cannot eject photoelectrons
from copper**.

   *Before the discussion on Compton scattering, it is important to understand
"X-rays."*

2.7 X-RAYS

In 1895, W. Röntgen discovered that "rays" from the tube could pass through
materials that were opaque to light and activate a fluorescent screen or
photographic film. Röntgen investigated and discovered that all materials were
transparent to them to some degree and that the degree of transparency decreased
with increasing density of the material. He was the first recipient of the Nobel
Prize for physics in 1901.

So, what are **X-rays**? X-rays are emitted when electrons are decelerated by crashing into a target in an X-ray tube. An X-ray spectrum consists of a series of sharp lines called the characteristic spectrum superimposed on the continuous bremsstrahlung spectrum (from the German for "braking radiation"). The minimum wavelength in the bremsstrahlung spectrum λ_m corresponds to the maximum energy of the emitted photon, which equals the maximum KE of the electrons, eV, where V is the voltage of the X-ray tube. Thus, the minimum wavelength is then given by $\lambda_m = \dfrac{hc}{eV}$.

Example 2.9

What is the minimum wavelength of the X-rays emitted by a television picture tube with a voltage of 2000 V?

Solution

The maximum KE of the electrons is 2000 eV, so this will be the maximum energy of the photons in the X-ray spectrum. The wavelength of a photon of this energy is the cutoff wavelength, which is $\lambda_m = \dfrac{hc}{eV} = \dfrac{1240}{2000} = \mathbf{0.62\ nm}$.

Example 2.10

Show that for X-rays, the wavelength λ can be given by $\lambda = 12.40/E$, where E is the energy in keV and λ is in Å (note: 1 nm = 10 Å).

Solution

We know that hc = 1240 eV nm. Thus, by unit manipulation, 1240 eV nm $\times$ $\dfrac{1\ keV}{1000\ eV} \times \dfrac{10\ Å}{1\ nm} = 12.40$ keVÅ. Thus, the proof is completed.

2.8 COMPTON SCATTERING

Arthur H. Compton provided information to further correct the photon concept. He measured the scattering of X-rays by free electrons. According to classical theory, when an electromagnetic wave of frequency n_1 is incident on material containing charges, the charges will oscillate with this frequency and will reradiate electromagnetic waves of the same frequency. Compton indicated that if this interaction were described as a scattering process involving a collision between a photon and an electron, the electron would recoil and thus absorb energy. That is, the scattered photon would have less energy and therefore a lower frequency than the incident photon.

Here, the energy and momentum of an electromagnetic wave can be related by E = pc. This result is consistent with the relativistic expression relating the energy and momentum of a particle if the mass of the photon is assumed to be zero.

From Figure 2.2, a photon from the incident X-rays collides with an electron initially at rest. Compton related the scattering angle θ to the incident and

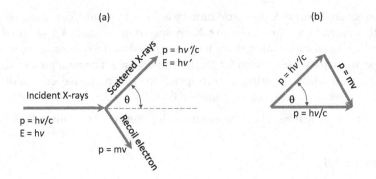

FIGURE 2.2 Illustration of (a) scattering and (b) conservation of momentum.

scattered wavelengths by treating the scattering as a relativistic mechanics problem and using the conservation of energy and momentum. Compton's result is $\lambda_2 - \lambda_1 = \dfrac{h}{mc}(1 - \cos\theta)$, where λ_1 and λ_2 are incident and scattered wavelengths, respectively. With the photon relations $\lambda = c/v$ and $E = hv$, $\dfrac{1}{E_2} - \dfrac{1}{E_1} = \dfrac{1}{mc^2}(1 - \cos\theta)$.

We can observe here that the change in wavelengths is independent of the original wavelength. Furthermore, the quantity h/mc depends only on the mass of the electron and has the dimension of length, called the **Compton wavelength** (λ_c). Its value is $\lambda_c = h/mc = 2.43 \times 10^{-12}$ m = 2.43 pm. Compton's experimental results for $\lambda_2 - \lambda_1$ as a function of the scattering angle θ agreed with the above equation, thereby confirming the correctness of the photon concept. *So, the factor* h/mc, *known as the* Compton wavelength, *is the wavelength shift that occurs when* $\theta = 90°$.

Example 2.11

Calculate the percentage change in wavelength observed in the Compton scattering of 20 keV photons at $\theta = 60°$.

Solution

The change in wavelength at 60° is given by $\lambda_2 - \lambda_1 = \lambda_c(1 - \cos\theta) = (2.43 \text{ pm})$ $(1 - \cos 60°) = 1.22$ pm. The wavelength of the incident 20-keV photons is $\lambda_1 = \dfrac{1240 \text{ eV} \times \text{nm}}{20,000 \text{ eV}} = 0.062$ nm = 62 pm. Thus, the percentage change in wavelength is $\dfrac{\lambda_2 - \lambda_1}{\lambda_1} = \dfrac{1.22}{62} \times 100\% = 1.97\%$.

Example 2.12

Show that the Compton wavelength evaluated for scattering from electrons is 2.43 pm.

Solution

Inserting the rest masses of the electron, which is 9.1×10^{-31} kg, into h/mc yields the Compton wavelengths of $\dfrac{6.63 \times 10^{-34} \text{ Js}}{\left(9.1 \times 10^{-31} \text{ kg}\right)\left(3 \times 10^{8} \text{ m/s}\right)} = 2.43 \times 10^{-12}$ m = **2.43 pm.**

Example 2.13

A photon with wavelength of 0.400 nm strikes an electron at rest and rebounds at an angle of 120° to its original direction. Determine the speed and wavelength of the photon after the collision.

Solution

We know that the speed of a photon is always the *speed of light in vacuum, c*. To obtain the wavelength after collision, we must use the equation for the Compton

$$\lambda_2 = \lambda_1 + \lambda_c (1 - \cos\theta) = 4 \times 10^{-10} \text{ m} + \left(2.43 \times 10^{-12} \text{ m}\right)\left(1 - \cos 120°\right)$$

effect: $= 4 \times 10^{-10} \text{ m} + \left(2.43 \times 10^{-12} \text{ m}\right)\left(1 + 0.5\right) = 4.036 \times 10^{-10} \text{ m}$

$$= \mathbf{0.4036 \text{ nm}.}$$

Let's work on a short revision exercise.

PROBLEM 2.4

Determine the recoil kinetic energy (MeV) of the electron that scatters 4 MeV photons by 50°?

PROBLEM 2.5

A photon with $\lambda = 0.5$ nm strikes a free electron head-on and is scattered straight backward. If the electron is initially at rest, what is its speed (km/s) after the collision?
When you have finished, move on to check your answers.
Answers:
Problem 2.4: **2.947 MeV**
Problem 2.5: **2900 km/s**
Here are the results in detail.

SOLUTION TO PROBLEM 2.4

We must note that $\dfrac{1}{E_2} - \dfrac{1}{E_1} = \dfrac{1}{mc^2}(1 - \cos\theta)$, where E_1 and E_2 are the incident and scattered photon energies, respectively. Therefore, with the given information, we first can calculate the energy of the scattered photon using this equation appropriately. First, we can calculate the mc^2 for the electron, which is $(9.1 \times 10^{-31} \text{kg})$ $(3 \times 10^8 \text{m/s})^2 = 8.19 \times 10^{-14}$ J = 0.511 MeV (Note: 1 eV = 1.602×10^{-19} J). The incident photon has 4 MeV, and thus,

$$\frac{1}{E_2} = \frac{1}{E_1} + \frac{1}{mc^2}(1 - \cos\theta) = \frac{1}{4} + \frac{1}{0.511}(1 - 0.6427) = 0.9492 \frac{1}{\text{MeV}} \text{ and } E_2 = 1.053 \text{MeV}.$$

Due to the conservation of energy, the kinetic energy of the recoiling electron must equal the energy lost by the photon. That is, $KE_{Re} = 4 - 1.053 =$ **2.947 MeV**.

SOLUTION TO PROBLEM 2.5

By the conservation of momentum, this gives $\dfrac{h}{\lambda} = m_e v - \dfrac{h}{\lambda'}$. Assume that the scattered electron is nonrelativistic. With the given information, back-scattered ($\theta = 180°$), this implies that $\lambda' - \lambda = \dfrac{h}{m_e c}(1 - \cos 180°) = \dfrac{2h}{m_e c}$. And $\lambda' = \lambda + \dfrac{2h}{m_e c} = 0.5 + 0.0048 = 0.5048$ nm. So, the electron velocity after collision is

$$v = \frac{h}{m_e}\left(\frac{1}{\lambda} + \frac{1}{\lambda'}\right) = \frac{6.63 \times 10^{-34}}{9.1 \times 10^{-31}}\left(3.98 \times 10^9\right) = 2.9 \times \frac{10^6 \text{ m}}{\text{s}} = \textbf{2900 km/s.}$$

2.9   ELECTRON WAVES → QUANTUM THEORY

In 1924, L. de Broglie suggested in his dissertation that electrons may have wave properties. It was based on the symmetry of nature. Since light was known to have both wave and particle properties, perhaps matter—especially electrons—might also have both wave and particle characteristics. It was speculative during that time due to prove of evidence. For the v and λ of electron waves, de Broglie chose the following equations:

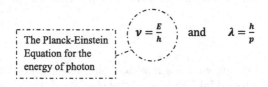

The Planck-Einstein Equation for the energy of photon $\qquad v = \dfrac{E}{h} \qquad$ and $\qquad \lambda = \dfrac{h}{p}$

$$v = \frac{E}{h} \text{ and } \lambda = \frac{h}{p}.$$

The wavelength equation also holds for photons; that is, $\lambda = \dfrac{c}{v} = \dfrac{hc}{hv} = \dfrac{hc}{E}$. Since the momentum of a photon is related to its energy by $E = pc$, then it should become

$$\lambda = \frac{hc}{pc} = \frac{h}{p}.$$

This is known as the **de Broglie wavelength**. De Broglie's equations are thought to apply to all matter. However, for macroscopic objects, the wavelengths calculated from $\lambda = h/p$ are so small that it is impossible to observe the usual wave properties of interference or diffraction. Even a particle as small as 1 μg is much too massive for any wave characteristics to be noticed, as we will see in this example.

Example 2.14

Find the de Broglie wavelength of a particle of mass 10^{-6}g moving with a speed of 10^{-6} m/s.

Solution

Here, we can use the formula, $\lambda = h/p$. Then, we would have

$$\lambda = \frac{h}{p} = \frac{h}{mv} = \frac{6.63 \times 10^{-34} \text{ Js}}{\left(10^{-9} \text{ kg}\right)\left(10^{-6} \text{ m/s}\right)} = 6.63 \times 10^{-19} \text{ m}.$$

Since the wavelength found in this example is much smaller than any possible apertures or obstacles (the diameter of the nucleus of an atom is about 10^{-15} m, roughly 10,000 times this wavelength), diffraction or interference of such waves cannot be observed. Here, the propagation of waves of very small wavelength is indistinguishable from the propagation of particles. In addition, the momentum is extremely small. A macroscopic particle with a greater momentum would have an even smaller de Broglie wavelength. Thus, we do not observe the wave properties of such macroscopic objects as baseballs or billiard balls.

2.10   BASIC QUANTUM THEORY

The wave nature of electrons was observed experimentally first by Davisson and Germer and later by G.P. Thomson, who measured the diffraction and interference of electrons. The mathematical theory of the wave nature of matter is known as **quantum theory**. In this theory, the electron is described by a wave function that obeys a wave equation. Energy quantization arises from standing wave conditions applied to electrons in various systems. Quantum theory is the basis for our understanding of the physical nature of the modern world.

2.11   EIGHT IMPORTANT POINTS IN QUANTUM MECHANICS

This topic, in general, can take the entire semester to fully cover. We only explore the general concepts to convey few points in wave mechanics in order to explain the general behavior of atomic electrons.

Point #1

- The state of a particle, such as an electron, is described by its wave function Ψ which is the solution of the **Schrödinger wave equation**.
- The absolute square of the wave function $|\Psi|^2$ measures the probability of finding the particle in some region of space.

Point #2

- A harmonic wave of a single angular frequency ω and wave number k can represent an electron that is completely non-localized and can be anywhere in space.

- In terms of the angular frequency and wave number, the de Broglie equations are $E = h\omega$ and $p = hk$ where $h = \dfrac{h}{2\pi}$.
- A localized electron can be represented by wave packet, which is a group of waves of nearly equal frequencies and wavelengths.
- The wave packet moves with a group velocity (V_g): $V_g = \dfrac{d\omega}{dk}$, which equals the velocity of the electron.

Point #3—**Wave–Particle Duality**

- Light, electrons, neutrons, and all other carriers of momentum and energy exhibit both wave and particle properties.
- Everything propagates like a wave, exhibiting diffraction and interference, and exchanges energy in discrete lumps like a particle.
- Because the wavelengths of macroscopic objects are so small, diffraction and interference are not observed.
- Also, when a macroscopic amount of energy is exchanged, so many quanta are involved that the particle nature of the energy is not evident.

Point #4—**The Uncertainty Principle**

- Wave–Particle Duality leads to the uncertainty principle, which states that the product of the uncertainty in a measurement of the position of a particle and the uncertainty in a measurement of its momentum must be greater than $h/4\pi$; that is, $\Delta x\,\Delta p \geq \dfrac{h}{4\pi}$.
- This was first enunciated by Werner Heisenberg in 1927.
- Similarly, the uncertainty in the energy ΔE is related to the time interval Δt required to measure the energy by $\Delta E\,\Delta t \geq \dfrac{h}{4\pi}$.
- An important consequence of the uncertainty principle is that a particle confined in space has a minimum energy called the zero-point energy.

Point #5—**Schrödinger equation**

- The wave function $\Psi(x, t)$ obeys the time-dependent Schrödinger equation:
$$-\frac{h^2}{8\pi^2 m}\nabla^2\Psi(x) + V(x)\Psi(x,t) = i\frac{h}{2\pi}\nabla\Psi(x,t).$$
- For any wave function that describes a particle in a state of definite energy, the time-dependent Schrödinger equation can be simplified by writing the wave function in the form $\Psi(x,t) = \psi(x)e^{-i\omega t}$.
- This leads to the time-independent Schrödinger equation (for one dimension): $-\dfrac{h^2}{8\pi^2 m}\dfrac{d^2\psi(x)}{dx^2} + V(x)\psi(x) = E\psi(x)$

- In addition to satisfying the Schrödinger equation, a wave function $\Psi(x)$ must be continuous and must have a continuous first derivation $d\Psi/dx$.
- Since the probability of finding an electron somewhere must be 1, the wave function must obey the normalization condition $\int_{-\infty}^{\infty} |\psi|^2 \, dx = 1$.
- This condition implies the boundary condition that Ψ must approach 0 as x approaches $\pm\infty$. Such boundary conditions lead to the quantization of energy.

Point #6

- An electron in a stationary state can be pictured as a cloud of charge with charge density proportional to $|\Psi|^2$.

Point #7

- When the quantum numbers of a system are very large, quantum calculations and classical calculations agree—a result known as Bohr's correspondence principle.

Point #8

- A wave function that describes two identical particles must be either symmetric or antisymmetric when the coordinates of the particles are exchanged.
- Fermions, which include electrons, protons, and neutrons, are described by antisymmetric wave functions and obey the Pauli exclusion principle which states that no two particles can have the same quantum number.
- Bosons, which include a particles, deuterons, photons, and mesons, have symmetric wave functions and do not obey the Pauli exclusion principle.

We will explore more examples here to illustrate these concepts.

Example 2.15

Find the de Broglie wavelength of a thermal neutron of mass 1.67×10^{-27} kg traveling at a speed of 2200 m/s.

Solution

We must use the de Broglie wave equation, $\lambda = h/(mv)$ where m is the mass of neutron in this case. Thus, $\lambda = \dfrac{6.63 \times 10^{-34} \text{ Js}}{(1.67 \times 10^{-27} \text{ kg})(2200 \text{ m/s})} = 1.8 \times 10^{-10} \text{ m} = 0.18 \text{ nm.}$

Example 2.16

Determine the de Broglie wavelength for a particle moving with the speed of 2000 km/s if the particle is (a) a proton and (b) a 0.1 kg ball.

Solution

We will make use of the de Broglie wave equation again,

$$\lambda = \frac{h}{m_{object}v} = \frac{6.63\times10^{-34}\text{ Js}}{m_{object}\left(2\times10^{6}\text{ m/s}\right)} = \frac{3.3\times10^{-40}}{m_{object}}\ [\text{Note: unit is meter}].$$

Substituting the value for m_{object}, one would find that the wavelength is **2×10^{-13}m for the proton, and 3.3×10^{-39}m for the 0.1 kg ball**. Notice that the de Broglie wavelength can be negligible for the macroscopic system indeed.

Example 2.17

If the uncertainty in the time during which an electron remains in an excited state is 10^{-7}s, what is the least uncertainty (in J) in the energy of the excited state?

Solution

Let E be the energy of the excited state. Then, $\Delta E \Delta t \geq \dfrac{h}{4\pi} \Rightarrow (\Delta E)(10^{-7}) \geq$

$\dfrac{6.63\times10^{-34}}{4\pi} \Rightarrow \Delta E \geq 0.528\times10^{-27}\,\text{J}.$

Example 2.18

By assuming that the uncertainty in the position of a particle is equal to its de Broglie wavelength, determine the relationship between the uncertainty in its velocity and its velocity.

Solution

We must use the Heisenberg uncertainty principle with the fact that $\Delta x = \lambda = h/(mv)$. Thus, $\Delta x\Delta(mv) \geq \dfrac{h}{4\pi} \rightarrow \dfrac{h}{mv}\Delta(mv) \geq \dfrac{h}{4\pi}$. Since m is constant, $\dfrac{h}{mv}\Delta(mv) \geq \dfrac{h}{4\pi}$

$\rightarrow \dfrac{h}{v}\Delta v \geq \dfrac{h}{4\pi} \rightarrow \Delta v \geq \dfrac{1}{4\pi}v$. The expression uncertainty of its velocity is $1/4\pi$ of its velocity indeed.

Now by way of revision, work out the following problem.

PROBLEM 2.6

At what energy (MeV) will the nonrelativistic calculation of the de Broglie wavelength of an electron be in error by 5%?

SOLUTION

The de Broglie wavelength for the nonrelativistic case is $\lambda_{nr} = \dfrac{hc}{pc} = \dfrac{hc}{\sqrt{2m_0c^2KE}}$.

Now, for the relativistic case, $(KE + m_0c^2)^2 = (pc)^2 + (m_0c^2)^2$, and it is possible to solve for the expression of "pc" such that the de Broglie wavelength is

$\lambda = \dfrac{hc}{pc} = \dfrac{hc}{\sqrt{2m_0c^2KE\left(1+\dfrac{KE}{2m_0c^2}\right)}}$. In this case $\lambda_{nr} - \lambda = 0.05\lambda$; that is, $\lambda_{nr}/\lambda = 1.05$.

So, $\dfrac{\lambda_{nr}}{\lambda} = \sqrt{1 + \dfrac{KE}{2m_0c^2}} = 1.05$ and $1.05 = \sqrt{1 + \dfrac{KE}{2(0.511\text{ MeV})}}$. One can now solve for

KE and show that the kinetic energy associating with this error is **KE = 0.105 MeV**.

Important Terms (in the Order of Appearance)

Photoelectric effect	De Broglie wavelength
Compton scattering	Quantum theory
Photon	Schrödinger wave equation
Work function	Wave–particle duality
Einstein's photoelectric function	The uncertainty principle
X-ray	Pauli exclusion principle
Compton wavelength	

BIBLIOGRAPHY

Bush, H.D., *Atomic and Nuclear Physics,* Prentice Hall, Englewood Cliffs, NJ, 1962.
Mayo, R.M., *Nuclear Concepts for Engineers,* American Nuclear Society, La Grange Park, IL, 1998.
Shultis, J.K. and R.E. Faw, *Fundamentals of Nuclear Science and Engineering 2nd Ed.,* CRC Press Taylor & Francis Group, Boca Raton, FL, 2008.
Shultis, J.K. and R.E. Faw, *Radiation Shielding,* American Nuclear Society, La Grange Park, IL, 2000.
Tipler, P.A., *Physics for Scientists and Engineers,* Worth Publishers, New York, NY, 1976.

FURTHER EXERCISES

A. True or False: If the statement is false, give a counterexample or explain the correction. If the statement is true, explain why it is true.
1. Wave–particle duality leads to the uncertainty principle.
2. Schrödinger's wave equation is the fundamental equation of quantum mechanics.
3. A photon's wavelength can be expressed as $\lambda = (1.24\times10^{-6}\text{m})/E$ with E being energy in eV.
4. The de Broglie wavelength of a 50 g golf ball traveling at 50 m/s is 2.65×10^{-30} m.
5. X-rays are emitted when electrons are decelerated by crashing into a target in an X-ray tube.
6. Photoelectric effects occur when photon collides with a free electron and emerges with reduced energy.
7. The energy in electromagnetic radiation is the product of Planck's constant and the frequency.
8. The energy of the photons in a beam whose wavelength is 526 nm is about 2.8 eV.

9. Pauli's exclusion principle states that no three particles can have the same quantum number.
10. An important consequence of the uncertainty principle is that a particle confined in space has a minimum energy called the zero-point energy.

B. Problems:
   1. How much energy (MeV) must be given to an electron to accelerate it to 95% of the speed of light?
   2. If 0.001 g of matter could be converted entirely into energy, what would be the value ($) of the energy so produced, at 1 cent per kWh?
   3. Determine the cutoff wavelength (nm) of X-rays produced by 100 keV electrons in a Coolidge tube.
   4. Determine gamma ray energy (MeV) with a wavelength of 1.25×10^{-12} m.
   5. Determine the maximum KE of photoelectrons ejected from a potassium surface by ultraviolet light ($\lambda = 200$ nm). Note: The photoelectric threshold wavelength for potassium is 440 nm. What retarding potential difference is required to stop the emission of these electrons?
   6. A photon of energy 5.0 eV imparts all its energy to an electron that leaves a special modified alloy surface with 1.0 eV of kinetic energy. Determine the work function (eV) of this modified alloy.
   7. Discuss about the general concept of "Wave–Particle Duality."
   8. An electron is confined to a region of space of length $L = 0.1$ nm. By assuming that the momentum p is equal to its change, $p = \Delta p$, then calculate
      a. The minimum uncertainty in its momentum (kg m s^{-1}),
      b. Its speed (m s^{-1}),
      c. The ratio of its relativistic mass to its rest mass, m/m_0.
   9. Suppose that a beam of 0.001 MeV photons is scattered by the electrons in an iron target.
      a. Determine the wavelength (nm) associated with these photons.
      b. Determine the wavelength (nm) of those photons scattered through an angle of 45 degrees.
      c. Determine the energy (MeV) of the scattered photons that emerge at an angle of 60° relative to the incident direction.
   10. Determine the de Broglie wavelength (m) of a proton accelerating through a potential difference of 1000 V.

3 Atomic and Nuclear Models Including Chart of Nuclides

OBJECTIVES

After studying this chapter, the reader should be able to:

1. Gain basic knowledge of Becquerel's discovery on radioactivity.
2. Learn about Thomson's atomic model and Rutherford atomic model.
3. State Bohr's postulates and describe the Bohr model of the hydrogen atom.
4. Sketch an energy-level diagram for hydrogen, indicate on its transitions involving the emission of a photon, and use it to calculate the wavelengths and energy of the emitted photons.
5. Explain about the model of nucleus and its discovery from the proton–electron model to proton–neutron model.
6. State about the nuclear stability based on nuclear shell model, discuss about the magic numbers associating with the band of stability, and predict the relative stabilities of nuclides.
7. Understand about the liquid drop model and basic binding energy terms associated with the proposed formulation.
8. Gain basic ideas on how to use the Chart of the Nuclides to identify different nuclides found in nature.

3.1 BASIC HISTORY

It is important to look into the historical timeline prior to going in detail of each discovery.

Year	Scientist	Discovery
1774	Antoine Lavoisier	Air = nitrogen + oxygen
1775	Henry Cavendish	Water = hydrogen + oxygen
1830s	Michael Faraday & Others	Basic of electricity
1854	Heinrich Geissler	Cathode ray tube
1860s	Dmitri Mendeleev	Periodic table
1874	G. Johnstone Stoney	Named the electron
1895	Wilhelm Röntgen	X-rays
1896	Henri Becquerel	Radiation
1897	J.J. Thomson	Cathode rays

DOI: 10.1201/9781003272588-3

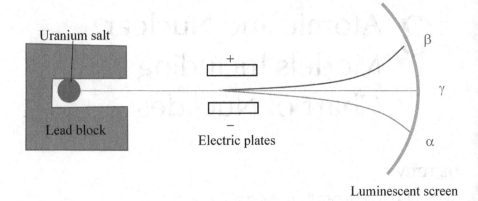

FIGURE 3.1   Illustration of the discovery of alpha, beta, and gamma rays.

It should be noted that during the 1800s, electricity was the main focus (simple experiments using electricity to show that atoms could be ionized). This is the beginning
of the journey that we will take in this chapter. We will explore here a brief historical
summary of the development of atomic and nuclear models. The readers should refer
to any modern physics textbook for more information.

3.2 DISCOVERY OF RADIOACTIVITY

In 1896, Becquerel discovered that uranium salts emitted rays similar to X-rays. Using
a magnetic field and a photographic plate, three types of radiation were identified—
alpha (α), beta (β), and gamma (γ) rays, as shown in Figure 3.1.

It can be observed that (1) alpha rays bend away from a positive plate and toward
a negative plate, indicating that they have a positive charge (known to consist of
helium-4 nuclei [nuclei with two protons and two neutrons]), (2) beta rays go opposite
direction showing that they have a negative charge (known to consist of high-speed
electrons), and (3) gamma rays are not being affected by electric and magnetic fields.
Uranium salts contain a number of radioactive elements, each emitting one or more
of these radiations.

We will discuss more in next chapter.

3.3 THOMSON'S ATOMIC MODEL: THE PLUM PUDDING MODEL

J.J. Thomson in 1897 demonstrated that the rays of a cathode ray tube could be
deflected by electric and magnetic fields. Under various combinations of electric and
magnetic fields, Thomson showed that all the particles had the same charge-to-mass
ratio, q/m. He tested with many materials for the cathode to show the q/m concept
indicating that these particles, so-called "electrons," were a fundamental constituent
of all matter (Figure 3.2).

Results of Becquerel and Thomson's experiments showed that atoms were transformed by emitting positively or negatively charged particles. This led to the first

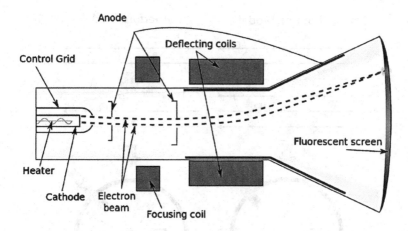

FIGURE 3.2   Illustration of the Thomson's experiment. (Ref: https://commons.wikimedia. org/wiki/File:Cathode_ray_tube_diagram-en.svg.)

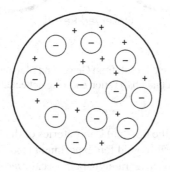

FIGURE 3.3   Plum pudding model based on Thomson's theory.

theories of atomic structure that atoms were composed of positively and negatively charged particles. During that time, there was no research information on the distributions of positive and negative charges in the atom. Thus, Thomson proposed a model by assuming that an atom consisted of a sphere of positive charge of uniform density with equal distributions of negative charge in the form of electrons. The atom was being viewed as a "**plum pudding**" with *dispersion of electrons like raisins or plums in a spherical dough of positive charges* (see Figure 3.3). During that time, Thomson could not find a possible configuration of electrons that would yield frequencies that are in agreement with the measured frequencies of the spectrum of any atom based on his model.

3.4   THE RUTHERFORD ATOMIC MODEL

The plum pudding model suggested by Thomson was later ruled out by a set of experiments performed by H.W. Geiger and E. Marsden under a supervision of E.

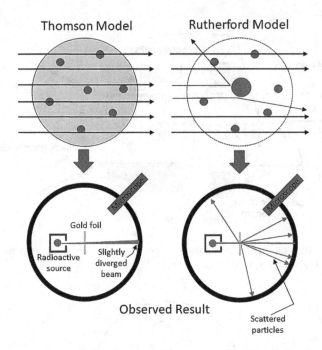

FIGURE 3.4 Left: The expected result from Geiger and Marsden based on Thomson's plum pudding model. Right: The actual observation showing deflected alpha particles.

Rutherford around 1911 at the Physical Laboratories of the University of Manchester; this is known as the Rutherford gold foil experiment (see Figure 3.4). They observed alpha particles from radioactive radium being scattered by atoms in a gold foil. Rutherford reported that the number of alpha particles scattered at large angles could not be coming from the well-distributed positive charge within that atomic volume (~0.1 nm in diameter). The results indicated that the positive charge and most of the mass of the atom must be centralized in a small region (i.e., the nucleus), with a diameter of the order of 10^{-15} m or 1 fm.[1]

3.5 THE BOHR ATOMIC MODEL

Niels Bohr, who was working in the Rutherford laboratory at that time, proposed a model of the hydrogen atom that combined the work of Planck, Einstein, and Rutherford to show a successful prediction of the observed spectra. Bohr proposed that the electron in the hydrogen atom revolved around the positive nucleus under the influence of the Coulomb attraction. For simplicity, he chose a circular orbit obeying the law of classical mechanics. To overcome classical electromagnetic theory, Bohr modified the laws of electromagnetism and *postulated* that *the electron could only move in certain nonradiating orbits,* known as "**Bohr's first postulate.**" He referred to these stable orbits as **stationary states**. The atom radiates only when the electron makes a transition from one stationary state to another. **Bohr's second postulate** is *that photon frequency v is related to the energies of the orbits* by

$$\nu = \frac{E_i - E_f}{h},$$

where h is Planck's constant, and E_i and E_f are the total energies in the initial and final orbits, respectively. This is an important key postulation in the Bohr theory because it differs from classical theory, which requires the frequency of radiation to be that of the motion of the charged particle. Next, let's explore the mathematical aspect of this model.

3.6 MATHEMATICAL DERIVATION OF THE BOHR ATOMIC MODEL AND BOHR'S THIRD POSTULATE

Figure 3.5(A) shows an electron of charge −e traveling in a circular orbit of radius r around the nuclear charge +Ze for the hydrogen atom. Thus, the potential energy PE at a distance r is

$$PE = -\frac{kZe^2}{r},$$

where k is the Coulomb constant.[2] Thus, the total energy E of the electron moving in a circular orbit with a speed v is $E = \frac{1}{2}mv^2 + PE = \frac{1}{2}mv^2 - \frac{kZe^2}{r}$. The kinetic energy can be obtained by using Newton's second law. We can set the Coulomb attractive force equal to the mass times the centripetal acceleration; thus, this becomes

$$\frac{kZe^2}{r^2} = m\frac{v^2}{r} \text{ or } \frac{1}{2}mv^2 = \frac{1}{2}\frac{kZe^2}{r}.$$

The total energy can be expressed as a function of r

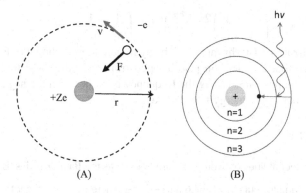

(A)   (B)

FIGURE 3.5 (A) Atomic model and (B) Bohr model of the hydrogen atom.

$$E = \frac{1}{2}mv^2 + PE = \frac{1}{2} \times \frac{kZe^2}{r} - \frac{kZe^2}{r} = -\frac{1}{2} \times \frac{kZe^2}{r}.$$

Now, we can apply the idea of the frequency of radiation emitted when the electron changes from orbit 1 of radius r_1 to orbit 2 of radius r_2 to get

$$v = \frac{E_1 - E_2}{h} = \frac{1}{2} \times \frac{kZe^2}{h}\left(\frac{1}{r_2} - \frac{1}{r_1}\right).$$

Bohr recognized that his derivation was analogous to the Rydberg–Ritz formula, $v = \frac{c}{\lambda} = cR\left(\frac{1}{n_2^2} - \frac{1}{n_1^2}\right)$, where R is the Rydberg constant $= 1.09677 \times 10^7 \mathrm{m}^{-1}$ and $n=1$, 2, 3,... at which $n_1 > n_2$. Bohr explored for a quantum condition for the radii of the stable orbits that would yield the squares of integers. He postulated that the angular momentum of the electron in a stable orbit equals an integer times Planck's constant divided by 2π. With the angular momentum of an allowable circular orbit equals mvr_n, this becomes $mvr_n = \frac{nh}{2\pi}$.

This is **Bohr's third postulate** for *quantizing angular momentum*. By solving for v and squaring it and balancing that with the orbital kinetic energy of the electron, one would get

$$v^2 = n^2 \frac{h^2}{4\pi^2 m^2 r_n^2} = \frac{kZe^2}{mr_n}.$$

Thus, the radius of an electron's orbit is $r_n = \frac{n^2 h^2}{4\pi^2 mkZe^2} = \frac{n^2 b_0}{Z}$, where $b_0 = \frac{h^2}{4\pi^2 mke^2} \approx 0.0529$ nm, is also called the first **Bohr radius**. The possible values of the energy based on this orbital radius is $E_n = -\frac{1}{2} \times \frac{kZe^2}{r_n} = -\frac{2\pi^2 k^2 Z^2 me^4}{n^2 h^2} = -Z^2\frac{E_0}{n^2}$, where $E_0 = \frac{2\pi^2 k^2 me^4}{h^2} \approx 13.6\,\mathrm{eV}$. The change in energy from one orbit n_1 to the second orbit n_2 is $\Delta E = \left(\frac{1}{n_1^2} - \frac{1}{n_2^2}\right)\frac{2\pi^2 k^2 Z^2 me^4}{h^2} = \left(\frac{1}{n_1^2} - \frac{1}{n_2^2}\right)Z^2 E_0$.

The visualization of Bohr model of the hydrogen atom is shown in Figure 3.5(B).

We can see that the electron just dropped from the third shell to the first with the emission of a photon showing the frequency is higher with the negative ΔE. If it jumps to a higher orbit, energy is absorbed.

Example 3.1

The set of spectral lines of hydrogen atom, known as the **Balmer series**, is described by the following formula: $\lambda = (364.6 \text{ nm})\frac{m^2}{m^2 - 4}$, where m is an integer that takes on the values of m=3, 4, 5, and so on. Balmer recommended that his formula might be a special case of a more general expression that would be applicable to the spectra of other elements. Show that the Balmer formula is indeed a special case of the Rydberg–Ritz formula, $v = \dfrac{c}{\lambda} = cR\left(\dfrac{1}{n_2^2} - \dfrac{1}{n_1^2}\right)$, for hydrogen ($Z=1$) with $n_2 = 2$ and $n_1 = m$.

Solution

If we take the reciprocal of the Balmer formula, then this would be

$$\frac{1}{\lambda} = \frac{1}{364.6 \text{ nm}}\left(\frac{m^2-4}{m^2}\right) = \frac{1}{364.6 \times 10^{-9} \text{ m}}\left(\frac{1}{1} - \frac{4}{m^2}\right)$$

$$= \frac{4}{364.6 \times 10^{-9} \text{ m}}\left(\frac{1}{4} - \frac{1}{m^2}\right) = \left(1.0970 \times 10^7 \text{ m}^{-1}\right)\left(\frac{1}{2^2} - \frac{1}{m^2}\right).$$

We can see that this is equivalent to the Rydberg–Ritz formula with R being close to the Rydberg constant. The interesting observation of this exercise is that n_2 in Rydberg–Ritz formula can be used to define a series of spectral lines. That is, by setting $n_2 = 1$, this would lead to a series in the ultraviolet region called the **Lyman series**, whereas setting $n_2 = 3$ leads to the **Paschen series** in the infrared region. Definitely, when $n_2 = 2$, this is the **Balmer series**.

Example 3.2

Show that the ionization energy of the hydrogen atom is about 13.6 eV.

Solution

We can estimate the ionization energy from the following equation:

$$E_0 = \frac{2\pi^2 k^2 m e^4}{h^2},$$

with $k = 8.987 \times 10^9$ N m^{2}C^{-2}, e = fundamental charge = 1.602×10^{19}C, m = electron mass = 511×10^3 eV/c^2, and $h = 4.135 \times 10^{-15}$ eV s. Thus, we have to balance the units (e.g., J [=] N m):

$$E_0 = \frac{2\pi^2 k^2 m e^4}{h^2} = \frac{2\pi^2 (8.987 \times 10^9 \text{ Nm}^2\text{C}^{-2})^2 \left(5.11 \times 10^5 \text{ eV/c}^2\right)(1.602 \times 10^{-19} \text{ C})^4}{(4.135 \times 10^{-15} \text{ eVs})^2}$$

$$= 19.739\left(8.0766 \times 10^{19} \text{ N}^2\text{m}^4\text{C}^{-4}\right)\left(5.11 \times 10^5 \text{ eV/c}^2\right)\left(6.58642 \times 10^{-76} \text{ C}^4\right)$$

$$\div \left(1.709 \times 10^{-29} \text{ eV}^2\text{s}^2\right)$$

$$= 3.1396 \times 10^{-20} / c^2 \text{N}^2\text{m}^4\text{eV}^{-1}\text{s}^{-2}$$

$$= 3.1396 \times 10^{-20} / \left(3 \times 10^8 \text{ ms}^{-1}\right)^2 \text{N}^2\text{m}^4\text{eV}^{-1}\text{s}^{-2}$$

$$= 3.4885 \times 10^{-37} \text{ N}^2\text{m}^2\text{eV}^{-1} \times \left(\frac{1 \text{ eV}}{1.602 \times 10^{-19} \text{ J}}\right)^2 = 13.59 \text{ eV} \approx \mathbf{13.6 \text{ eV}}.$$

The learning lesson here is the unit manipulation and careful arrangement of the units.

Example 3.3

The ionization energy of the hydrogen atom is measured to be 13.6 eV. Estimate the wavelength of the first line of the Lyman series.

Solution

The energy of a state of the hydrogen atom with quantum number n is given in frame 6 as

$$E_n = -\frac{1}{2} \times \frac{kZe^2}{r_n} = -\frac{2\pi^2 k^2 Z^2 me^4}{n^2 h^2} = -Z^2 \frac{E_0}{n^2} = -\frac{13.6}{n^2} \text{ eV},$$

where Z=1. We know from Example 3.1 that the first line of the Lyman series occurs from a transition from the n=2 state to the n=1 state. The energy of the photon is

$$E = E_2 - E_1 = -\frac{13.6}{2^2} + \frac{13.6}{1^2} = 10.2 \text{ eV}.$$

We can use the Planck's equation to find the wavelength, which is

$$\lambda = \frac{hc}{E} = (1240 \text{ eVnm})/(10.2 \text{ eV}) = \textbf{122 nm}.$$

Example 3.4

Calculate the first three energy levels (n=2, 3, and 4) of doubly ionized lithium and determine its ionization potential.

Solution

Lithium atom has three protons, Z=3, and the Bohr energy formula becomes

$$E_n = -Z^2 \frac{E_0}{n^2} = -\frac{13.6}{n^2}(3)^2 = -\frac{122.4}{n^2} \text{ eV}.$$

Thus, $E_1 = -122.4$ eV (ionization potential is 122.40 V), $E_2 = \textbf{-30.60 eV}$, $E_3 = \textbf{-13.60 eV}$, and $E_4 = \textbf{-7.65 eV}$.

It should be noted that the wavelengths will be $\frac{1}{9}$ of the corresponding wavelengths for hydrogen in each energy level, $\lambda_1 = 13.5$ nm, $\lambda_2 = 11.3$ nm, and $\lambda_3 = 10.8$ nm, respectively.

Now, let's have fun with these problems.

PROBLEM 3.1

Find the energy (eV) and wavelength (nm) of the line with the longest wavelength in the Lyman series for hydrogen atom.

PROBLEM 3.2

Determine the radiation of greatest wavelength (nm) that will ionize unexcited hydrogen atoms.

PROBLEM 3.3

Imagine that there is a way to strip 28 electrons from copper (Cu, $Z=29$) in a vapor of this metal. Calculate the first three energy levels for the remaining electron.
Turn to the next frame once you are done to check your answers.

Answers

Problem 3.1: **10.2 eV, 121.6 nm**
Problem 3.2: **91.2 nm**
Problem 3.3: $E_1 = -11.438$ keV; $E_2 = -2.859$ keV; $E_3 = -1.271$ keV
Detailed solutions are given below.

Solution to Problem 3.1

We know that $\Delta E = \left(\dfrac{1}{n_1^2} - \dfrac{1}{n_2^2}\right)\dfrac{2\pi^2 k^2 Z^2 me^4}{h^2} = \left(\dfrac{1}{n_1^2} - \dfrac{1}{n_2^2}\right) Z^2 E_0.$

And the Lyman series corresponds to transitions ending at the ground-state energy, $\Delta E = -13.6$ eV. You can substitute $n_2 = 1$ for Lyman series, $n_1 = \infty$, $Z = 1$, and $E_0 = 13.6$ eV to get that. Since the wavelength varies inversely with energy, the transition with the longest wavelength is the transition with the lowest energy, which is from the first excited state $n=2$ to the ground state $n=1$. The energy of the first excited state is $E_2 = (-13.6 \text{ eV})/4 = -3.40$ eV. Since this is 10.2 eV above the ground-state energy, the energy of the photon emitted is 10.2 eV and the corresponding wavelength of this photon is $\lambda = \dfrac{hc}{\Delta E} = (1240 \text{ eV nm})/(10.2 \text{ eV}) = \textbf{121.6 nm}$. It is important to note that this photon is outside the visible spectrum and in the ultraviolet region.

Solution to Problem 3.2:

We can see that $E_\infty - E_1 = 13.6$ eV; we can use this to determine the wavelength by setting: $E_\infty - E_1 = \dfrac{hc}{\lambda} = 13.6$ eV. Thus, $\lambda = 1240$ eV nm/13.6 eV $= \textbf{91.2 nm}$. Here, the wavelengths shorter than this would not only remove the electron from the atom but would give the removed electron KE. This is the shortest wavelength for a line in the Lyman series as well.

Solution to Problem 3.3

We can use the resulting observation from Example 3.4 to help rapidly solving this problem. Here, $Z=29$ and thus, the allowable energies will be $29^2 = 841$ based on the following formula: $E_n = -Z^2 \dfrac{E_0}{n^2}$. This is 841 times the corresponding energies for hydrogen; that is, $E_1 = 841(-13.60 \text{ eV}) = \textbf{-11.438 keV}$; $E_2 = E_1/4 = \textbf{-2.859 keV}$; and $E_3 = E_1/9 = \textbf{-1.271 keV}$.

3.7   FIRST-GENERATION MODEL OF NUCLEUS

More accurate measurements were made by scattering electron beams instead of alpha particles (after Rutherford's experiment). These experimental approaches led to formulas describing proton density inside spherical nuclei and then general nucleon density in spherical nuclei. By measuring the X-ray emission after capturing muon by a nucleus, the density of proton can be expressed as

$$\rho_{proton} = \frac{\rho_{proton}^0}{1 + \exp\left[\dfrac{r - R}{a}\right]} \left(\frac{protons}{fm^3}\right),$$

where r is the distance from the center of the nucleus, R is the radius of the nucleus at which the proton density falls to ½ of its central value, and a is the surface thickness over which the proton density becomes negligibly small. This is based on the equivalent between the total number of protons and the atomic number; that is,

$$4\pi \int_0^\infty r^2 \rho_{proton}(r)\,dr = Z.$$

If we assume that the density ratio of neutrons $\rho_{neutron}$ to protons ρ_{proton} inside the nucleon is represented by $N/Z = (A - Z)/Z$, then the nucleon density can be expressed as

$$\rho(r) = \rho_{proton}(r) + \rho_{neutron}(r) = \left(1 + \frac{A - Z}{Z}\right)\rho_{proton}(r) = \frac{A}{Z}\rho_{proton}(r) \left(\frac{nucleons}{fm^3}\right).$$

Thus, $\rho(r) = \dfrac{\dfrac{A}{Z}\rho_{proton}(r)}{1 + \exp\left[\dfrac{r - R}{a}\right]} = \dfrac{\rho^{\cdot}}{1 + \exp\left[\dfrac{r - R}{a}\right]} \left(\dfrac{nucleons}{fm^3}\right)$. The resulting equation shows that

the density at the center of the nucleus is 0.16×10^{45} nucleon/m^3 and the density at a distance R is equal to a multiple of $A^{1/3}$; that is, $R = 1.1\ A^{1/3} \times 10^{-15} m$.

Example 3.5

Based on the nucleon distribution given above, calculate the fraction the density of the nucleus decreases between $r = R - a$ and $r = R + a$.

Solution

We would find that $\rho(R - a) = \dfrac{\rho^*}{1 + \exp\left[\dfrac{(R-a) - R}{a}\right]} = \dfrac{\rho^*}{1 + \exp(-1)} = 0.731\rho^*$ and

$$\rho(R + a) = \dfrac{\rho^*}{1 + \exp\left[\dfrac{(R+a) - R}{a}\right]} = \dfrac{\rho^*}{1 + \exp(1)} = 0.269\rho^*.$$

Thus, the fraction that the nucleus decreases would be $0.269/0.731 = \mathbf{0.368}$. That is, the nuclear density decreases at R+a to 36.8% of its value at R − a.

Interestingly, each nucleon has the same volume because the total volume of the nucleus must be proportional to A through $(4/3)\pi R^3$. Each nucleon may have mass of 1.67×10^{-27} kg (mass of a proton in general for hydrogen). Then, the mass density will be $(0.16 \times 10^{45}$ nucleon/m$^3)$ $(1.67 \times 10^{-27}$ kg/nucleon$) = 2.67 \times 10^{17}$ kg/m$^3 = 2.67 \times 10^{14}$ g/cm^3. So what? Let's provide some perspective here. For example, density of lead is 11.4 g/cm^3 and density of uranium is 18.90 g/cm^3. This implies that a nucleus is 10 *trillion times* denser!

3.8 PROTON–ELECTRON MODEL

The above discovery and postulates have led us to many possible models to describe the nucleus and the forces associated to it to satisfy the volume and high density. Scientists have tried to find a proper concept to explain combinations of neutrons and protons that produce stable nuclei and unstable nuclei. The **proton–electron model** is depicted as an atom having a nucleus containing A protons and (A–Z) electrons with Z orbital electrons surrounding the nucleus. This is based on the idea that masses of all nuclei are observed to be whole numbers leading to the assumption that all heavy nuclei are composed of multiples of the hydrogen atom.

Example 3.6

If fluorine (F) is to follow the proton–electron model, explain the number of protons and electrons that it should have and its characteristic in that nature.

Solution

Fluorine (F) has an atomic mass number of 19 and an atomic charge of 9. Hence, the nucleus of a fluorine atom should contain **19 protons** to give it the correct mass and **10 electrons** to give it the correct charge. In addition, there must be 10 electrons surrounding the nucleus to produce a neutral fluorine atom.

However, this model encountered two major problems. First, the predicted angular momentum (or spin) of the nuclei did not match values found experimentally. Theoretically, both protons and electrons must have an inherent spin of $\frac{1}{2}(h/2\pi)$. The spin of the nucleus must therefore be a combination of its constituents spins. When the total spin is calculated, it must equal a whole number and this is not what was found experimentally.

Example 3.7

Show that there is a spin inconsistency for ^{9}Be.

Solution

Here, $^9_4$Be contains nine protons and five electrons for a total of 14 spins based on the proton–electron model. This should be $14\left(\pm\frac{1}{2}\right)$ resulting in 0 or 1. However, an experiment shows that the spin for the Be nucleus is $\frac{1}{2}$; therefore, there is a consistency in the spin.

The second concern is dealing with the Heisenberg uncertainty principle which indicates that the total energy for the electrons contained in the nucleus is 10 MeV. The rest mass of an electron is 0.511 MeV/c² and the energy of electrons being emitted from atoms is only a few MeV. Therefore, the energy predicted by the model is too high implying that there is a fundamental problem with having the electrons contained within the nucleus.

Example 3.8

Based on the proton–electron model, show that according to the Heisenberg uncertainty principle, the total energy for the electrons contained in the nucleus is about 10 MeV.

Solution

An electron must be confined within the nucleus diameter of ~10^{-14} m. Thus, its uncertainty in position is about $\Delta x = 10^{-14}$ m and the uncertainty in the electron's momentum Δp would be (information from Chapter 2): $\Delta p \Delta x \geq \dfrac{h}{4\pi}$.

It means that the minimum uncertainty in momentum of the electron should be

$$\Delta p \Delta x \geq \frac{h}{4\pi} \Rightarrow \Delta p \geq \frac{h}{4\pi \Delta x} = \frac{6.63 \times 10^{-34} \text{ Js}}{4\pi \left(10^{-14} \text{ m}\right)} = 5.276 \times 10^{-21} \text{ Jm}^{-1}\text{s}.$$

Energy of the electron can be calculated using the fact that (Chapter 2)

$$E = \sqrt{p^2 c^2 + m_0^2 c^4}$$

$$= \sqrt{(5.276 \times 10^{-21})^2 (3 \times 10^8)^2 \frac{1 \text{ eV}}{1.6 \times 10^{-19} \text{ J}} + (0.511 \text{ MeV}/c^2)^2 c^4}$$

$$= \mathbf{9.90 \text{ MeV}}.$$

This is the energy far greater than its rest-mass energy.

These two recognized problems indicated that there was a flaw within the proton–electron model, leading us to another model.

3.9 PROTON–NEUTRON MODEL

In 1932, Chadwick discovered neutrons which led Heisenberg to hypothesize that neutrons were located in the nucleus. This hypothesis solved both problems with the proton–electron model. Unfortunately, it too had problems: (1) repulsive Coulombic repulsive forces between positive charge of the protons and (2) no previous concept of a force strong enough to overcome the Coulombic repulsion. In the proton–neutron model, "nuclear force"—very strong (millions of eV) and short range (distance < nuclear radius)—has been proposed to be there to hold nucleons to each other. This is one of the major challenges in nuclear physics during that time to understand and quantify the nature of the nuclear force.

This new model idea leads us to discuss about the stability of nuclei, which will be discussed in the next frame.

3.10 NUCLEAR STABILITY

At first glance in a nucleus containing several protons, one important question that we might be asking would be "Should the protons be strongly repelling each other due to their like electric charges inside a nucleus?" Stable nuclei with more than one proton existed due to the **nuclear force**—*a strong force of attraction between nucleons that acts only at very short distances (about 10^{-15}m)*. Thus, two protons that are much farther apart than 10^{-15}m must repel one another. But, within the nucleus, two protons must be close enough for the nuclear force between them to be active. This force, in return, should be enough to compensate the repulsion of electric charges, yielding a stable nucleus.

The **nuclear shell model** is being considered under this developed conceptual design of the nucleus; it is *a nuclear model in which protons and neutrons exist in levels, or shells, analogous to the shell structure that exists for electrons in an atom.* We can look back to the filled shells of electrons associating with the special stability of the noble gases where stable atoms are 2 for He, 10 for Ne, 18 for Ar, and so forth. Experiments have shown that nuclei with certain numbers of protons or neutrons appear to be very stable. These numbers, called **magic numbers**, which can be explained using the shell model, are *the numbers of nuclear particles in a completed shell of protons or neutrons.* Because nuclear forces are different from electrical forces, these numbers are not the same as those for electrons in atoms. For protons, the magic numbers are 2, 8, 20, 28, 50, and 82, while neutrons have these same magic numbers, as well as the magic number 126. Additional calculation has shown that 114 should have been the magic number for protons.

Example 3.9

Discuss the special stability of $^4_2$He..

Solution

In many radioactive nuclei decay by emitting alpha particle $\left(^4_2\text{He}\right)$., there is a unique stability in this $^4_2$He nucleus because it contains two protons and two neutrons—a magic number of protons (2) and a magic number of neutrons (also 2).

Example 3.10

Is $^{208}_{82}$Pb a stable nuclide? Explain.

Solution

The answer is **"YES"** because it contains 82 protons, a magic number. Not only that, it has a magic number of neutron as well ($208 - 82 = 126$).

Experimental observations have shown the special stability of pairs of protons and pairs of neutrons to be analogous to the stability of pairs of electrons in molecules. The results shown in the below table indicate that only four stable isotopes have both odd numbers of protons and neutrons, whereas there are 159 stable isotopes with both even numbers of protons and neutrons. Overall, there are 266 stable nuclides.

	Number of Stable Isotopes			
Number of	159	53	50	4
Protons	Even	Even	Odd	Odd
Neutrons	Even	Odd	Even	Odd

Over 3200 known nuclides can be plotted on a graph with the number of protons on the horizontal axis and the number of neutrons on the vertical axis, as shown in Figure 3.6 The stable nuclides will fall in a certain band spread on the graph, known as the **band of stability**; this is *the region in which stable nuclides lie in a plot of number of protons against number of neutrons* (see Figure 3.6). The ratio of neutrons to protons is about 1.0–1.1 for nuclides up to $Z = 20$. As Z increases, this ratio increases to ~1.5 due to the increasing repulsions of protons from their electric charges. More neutrons are required to give attractive forces to offset these repulsions.

It can be observed experimentally as the number of protons becomes very large ($Z > 83$), the proton–proton repulsions would increase significantly that stable nuclides are impossible to exist; all nuclides with $Z > 83$ are radioactive. The nuclei

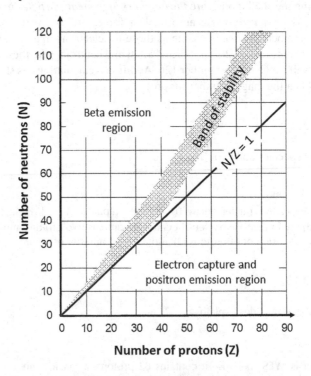

FIGURE 3.6 Band of stability showing stable nuclides.

of radioactive atoms undergo spontaneous change whereby the number of neutrons and/or protons change. These changes produce more stable nuclei. Emission of a single proton or neutron rarely occurs in nature, whereas emission of small nucleus containing two neutrons and two protons is more common. Mainly, an alpha particle $\left(_{2}^{4}He\right)$ is the key factor. Nuclei which are composed of multiples of alphas particles are extremely stable. These types of nuclei require an enormous amount of energy to break them apart—examples include ^{8}Be, ^{12}C, ^{16}O, and ^{20}Ne. On the contrary, all elements with Z equal to 83 or less have one or more stable nuclides, with the exception of technetium (Z=43) and promethium (Z=61).

Let's work on this problem:

PROBLEM 3.4

Which is radioactive and which is stable? Explain.
  (a) $_{84}^{210}Po$ (b) $_{83}^{209}Bi$ (c) $_{31}^{76}Ga$
If you have the answers, please move next.
  Here is the problem strategy and solution in the previous frame.

Solution to Problem 3.4

  a. Based on the general principles given in frame 13, polonium has an atomic number greater than 83 (i.e., Z=84). Therefore, $_{84}^{210}Po$ is **radioactive**.
  b. For bismuth, the given atomic mass is 209 and therefore, $209 - 83 = 126$ neutrons, which is a magic number. Hence, it is expected that $_{83}^{209}Bi$ is **stable**. *Note*: $_{83}^{209}Bi$ has a half-life of 19×10^{18} y, which is another evidence of a stable isotope. This will be discussed in the future chapter.
  c. Gallium is rather tricky because both Z (31) and N (45) are not magic numbers. In addition, stable odd–odd nuclei are rare, and $_{31}^{76}Ga$ lies farther from the center of the band of stability. For these two reasons, $_{31}^{76}Ga$ would be considered **radioactive**.
So, based on the above knowledge, $_{50}^{118}Sn$ is ......

3.11 LIQUID DROP MODEL

Liquid drop model is based on the concept that density of matter inside a nucleus is nearly the same for all nuclides and the volume of nucleus proportional to A. It suggests that forces holding nucleus together are saturated; that is, nucleon interacts only with its immediate neighbors. Its concept is similar to a liquid drop in which each molecules interacts only with its nearest neighbors. An almost constancy of density of nuclear matter is analogous to an incompressible liquid where the energy corrections were required to correlate the **binding energy**—the energy needed to break a nucleus into its individual protons and neutrons—predicted and that found experimentally. Although this model provides nothing about the internal structure of the nucleus, it provides a good predicting stable combination of N and Z numbers.
  The initial mathematical representation of this concept is

$$m\left(_{Z}^{A}X\right) = Zm_{p} + (A - Z)m_{n},$$

where $m\left(_Z^A X\right)$ is the nuclear mass of chemical element X composing of Z protons and $N = A-Z$ neutrons multiplying to its corresponding mass, m_p and m_n, respectively. Binding energy needs to be considered in the above equation to account for the conversion of a small fraction of the mass into energy when a nucleus is created. Wapstra in 1958 introduced the formulation with several binding energy adjustable factors necessary in balancing the above expression; these are:

1. Volume binding energy, BE_v, which should be proportional to the volume of the nucleus, namely A; that is, $BE_v = c_v A$, where c_v is a positive constant. This first factor often overestimates the total binding energy; the value is positive in nature.
2. Surface binding energy, BE_s, which is proportional to the surface area of the nucleus, $A^{2/3}$, describing the surface effect in interior nucleons. This second factor is expressed as $BE_s = -c_s A^{2/3}$, where c_s is a positive constant.
3. Coulombic binding energy, BE_c, which is proportional to the $Z^2 A^{-1/3}$, explaining the negative potential energy of the repulsive forces between protons inside the nucleus. This third factor is expressed as $BE_c = -c_c Z^2 A^{-1/3}$ where c_c is a positive constant.
4. Asymmetry binding energy, BE_a, which is expressed through an empirical form proportional to $(A-2Z)^2/A$, providing the symmetrical balance between neutrons and protons inside the nucleus. This fourth factor is expressed as $BE_a = -c_a(A-2Z)^2/A$, where c_a is a positive constant.
5. Pairing binding energy, BE_p, which can be expressed through an empirical form proportional to $A^{-1/2}$, describing the pairing proton–neutron sequence (discussed in frame 13) for the stability nuclides. This fifth factor is given by $BE_p = -c_p A^{-1/2}$, where c_p is positive for even–even nuclides, negative for odd–odd nuclides, and zero for even–odd or odd–even nuclides.

It should be noted that the second, third, fourth, and fifth factors can be considered as the correction binding energy terms. Combining these terms together, the mass of nucleus can be expressed as

$$m\left(_Z^A X\right) = Zm_p + (A-Z)m_n - \frac{1}{c^2}\left[c_v A - c_s A^{2/3} - c_c \frac{Z^2}{A^{1/3}} - c_a \frac{(A-2Z)^2}{A} - \frac{c_p}{\sqrt{A}} \right],$$

where $c_v = 15.835$ MeV, $c_s = 18.33$ MeV, $c_c = 0.714$ MeV, $c_a = 23.20$ MeV, and

$$c_p = \begin{cases} +11.2 \text{ MeV} & \text{for odd N and odd Z} \\ 0 & \text{for odd N, even Z or for even N, odd Z.} \\ -11.2 \text{ MeV} & \text{for even N and even Z} \end{cases}$$

Example 3.11

Calculate the mass of an atom of $_5^{18}B$ using the liquid drop model.

Solution

We know that the atomic mass $M(^{18}_{5}B) \approx m(^{18}_{5}B) + 5m_e$, if we neglect the binding energy of the electrons to the nucleus. This can be fully expressed as

$$M(^{18}_{5}B) = 5\,m_p + 13\,m_n - \frac{1}{931.5\,\text{MeV/u}}\left[c_v 18 - c_s 18^{2/3} - c_c \frac{5^2}{18^{1/3}} - c_a \frac{(18-10)^2}{18} - \frac{c_p}{\sqrt{18}} \right]$$

$+5\,m_e$.

We must recognize that both N and Z are odd and thus, $c_p = +11.2$ MeV in this case.

Z	5		c_v	15.835	MeV
A	18		c_s	18.33	MeV
N	13		c_c	0.714	MeV
m_e	0.000549	u	c_a	23.2	MeV
m_p	1.00727	u	c_p	11.2	MeV
m_n	1.008664	u			

	Volume	Surface	Coulombic	Asymmetry	Pairing	Total BE
(MeV)	285.03	−125.896	−6.8110497	−82.48889	−2.63987	67.19452
Total BE in u =		0.072136 u				

Thus, $M(^{18}_{5}B) \approx 5(1.00727\ u) + 13(1.008664\ u) - 0.072136\ u + 5(0.000549\ u)$ = 18.07959 u. The tabulated value from the open literature is 18.0517 u showing about 4.12% error in prediction.

Example 3.12

Find the percent difference between the actual atomic mass of ^{12}C and that using the liquid drop model.

Solution

We have to determine the atomic mass of ^{12}C using the liquid drop model. Note that both N and Z are even; thus, the value of $c_p = -11.2$ MeV.

Z	6		c_v	15.835	MeV
A	12		c_s	18.33	MeV
N	6		c_c	0.714	MeV
m_e	0.000549	u	c_a	23.2	MeV
m_p	1.00727	u	c_p	-11.2	MeV
m_n	1.008664	u			

	Volume	Surface	Coulombic	Asymmetry	Pairing	Total BE
(MeV)	190.02	−96.0764	−11.227256	0	3.233162	85.94953
Total BE in u		0.09227 u				

Here, $M(^{12}C) \approx 12.00662$ u, and the actual atomic mass is 12.00000 u; thus, the percent difference will be 0.0551%.

Let's work on something fun.

Example 3.13

Show that an expression for the line of stability from the liquid model is

$$Z(A) = \frac{A}{2}\left[\frac{1+\dfrac{(m_n - m_p)c^2}{4c_a}}{1+\dfrac{c_c A^{2/3}}{4c_a}}\right].$$

Did you get the same expression? If not, let's look at how to prove it.

Solution

The idea is to think about the relationship between N and Z that will produce the most stable nuclides. The hint is coming from the band of stability plot shown in frame 13. What you can see is that the most stable nuclide of the isobar (nuclides with the same A) is the one with the smallest mass. Therefore, we have to take a derivative of the liquid drop model expression with respect to Z at constant A:

$$\left[\frac{\partial m\left(^A_Z X\right)}{\partial Z}\right]_A = m_p - m_n - \frac{1}{c^2}\left[-2c_c\frac{Z}{A^{\frac{1}{3}}} + 4c_a\frac{(A-2Z)}{A}\right].$$

We then set this to zero and solve for Z at constant A to get

$$m_p - m_n - \frac{1}{c^2}\left[-2c_c\frac{Z}{A^{\frac{1}{3}}} + 4c_a\frac{(A-2Z)}{A}\right] = 0$$

$$-2c_c\frac{Z}{A^{\frac{1}{3}}} + 4c_a\frac{(A-2Z)}{A} = (m_p - m_n)c^2 \rightarrow -2c_c\frac{Z}{A^{\frac{1}{3}}} + 4c_a - \frac{8c_a Z}{A} = (m_p - m_n)c^2 \rightarrow$$

$$-\frac{2}{A}\left(c_c Z A^{\frac{2}{3}} + 4c_a Z\right) = (m_p - m_n)c^2 - 4c_a \rightarrow \left(c_c A^{\frac{2}{3}} + 4c_a\right)Z = \frac{A}{2}\left[(m_n - m_p)c^2 + 4c_a\right]$$

$$Z = \frac{A}{2}\frac{\left[(m_n - m_p)c^2 + 4c_a\right]}{c_c A^{\frac{2}{3}} + 4c_a} \rightarrow Z = \frac{A}{2}\left[\frac{1+\dfrac{(m_n - m_p)c^2}{4c_a}}{1+\dfrac{c_c A^{\frac{2}{3}}}{4c_a}}\right].$$

The value of Z versus N can be plotted on the band of stability figure. Amazingly, if you do it, you will see an excellent agreement with the observed nuclide stability trend.

We are now heading to the final topic in this chapter.

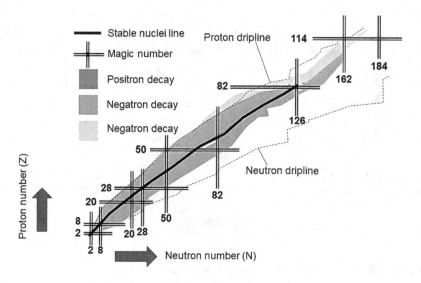

FIGURE 3.7   Representation of the Chart of Nuclides.

3.12   CHART OF THE NUCLIDES

There are over 3200 nuclides with a distinct combination of Z and A of which 266 nuclides are stable and are found in nature. Sixty-five long-lived radioisotopes are found in nature. The lightest atom is ordinary hydrogen (^{1}H), while the heaviest is continually increasing as it is being created/discovered in laboratories. These nuclides and their important properties have been displayed in a compact way on the **Chart of the Nuclides**—*a two dimensional arrays of atomic number Z (y-axis, 0–107) versus neutron number N (x-axis, 0–158).* General Electric Co. offers the most detailed Chart of the Nuclides and can be purchased. In addition, the automated Chart of the Nuclides can also be found and explored through a web-based site at http://www.nndc.bnl.gov/chart/; the representation of this chart is sketched and shown in Figure 3.7 and a partial portion of this chart is shown in Figure 3.8.

We are going to provide a basic step of how to use this chart.

Step 1: The chemical symbol, element name, and atomic mass are displayed for each chemical element in the first column of the chart (farthest left column of the chart).

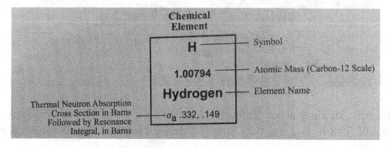

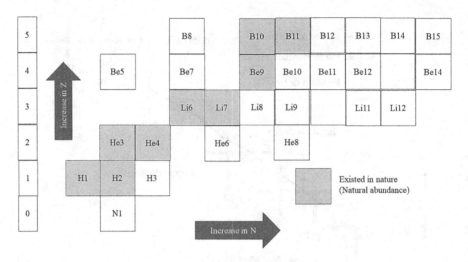

5			B8		B10	B11	B12	B13	B14	B15
4		Be5	Be7		Be9	Be10	Be11	Be12		Be14
3			Li6	Li7	Li8	Li9			Li11	Li12
2		He3	He4		He6		He8			
1	H1	H2	H3							
0		N1								

Increase in Z

Increase in N

Existed in nature
(Natural abundance)

FIGURE 3.8 Partial portion of the Chart of Nuclides.

Step 2: All the isotopes of that element are shown along the x-axis as shown
below. If it is radioactive, then **half-life**—*the time for each nuclide to reduce
its quantity to half its initial value*—is given (this will be discussed later). If
it is stable, the percentage of the element represented by that one nucleus or
isotope is shown below it. It is important to keep in mind that in each row,
all isotopes have the same chemical element with the same chemical behavior but with different atomic masses and nuclear behaviors.

		σ_a 3.5 mb, 1.6 mb	8.03768	E 16.498			B10	3+	B11	3/-
B 10.811			B7 (3/-) 3E-22 s	B8 2+ 770 ms	B9 3/- 8E-19 s		19.9* 384E1, 173E1		80.1*	
5			p, α	β+ 14 .. (α) 8.359 (2α) 1.57	p, 2α		α 3 γ 7 mb ε 8 mb		β_γ 5 mb, 2 mb	
Boron										
σ_a 76E1, −343			7.0299	E 17.979	9.013329		10.0129370		11.0093055	
Be 9.012182		Be6 5.0E-21 s		Be7 3/- 53.3 d	Be8 ~7E-17 s		Be9 3/- 100		Be10 1.5E6 a	
4		2p, α		ε	2α .0461		α, 8 mb, 4 mb		β− 556	
Beryllium				ε 477.6 γ 3.854, ~1.8E4 14, 0.06					β−, <1 mb	
σ_a 8 mb, 4 mb			8.01973	E 0.86182	8.00530509		9.0121821		E .556	
		Li4 2− 8E-23 s	Li5 3/- ~3E-22 s	Li6 1+ 7.59*	Li7 3/- 92.41*		Li8 2+ 0.840 s		Li9 3/- 178.3 ms	
		p	p, α	σα 941, 423 σγ 39 mb, 17 mb	σγ .045, .020		β− 13 (2α) 1.57		β− 13.5, 11.0,~ (n) 3,~ (2α) 7,~	
		4.0272	5.0125	6.015122	7.016004		E 16.0045		E 13.606	
		He3 1/+ 0.000137*	He4 1/+ 99.999863	He5 3/- 7.0E-22 s	He6 607 ms		He7 (3/)− 3E-21 s		He8 119 ms	
				n, α	β− 3.510 no γ (d) tot	n			β− 10.7 γ 980.7 (n) 81-3.0 (t) α	
		3.016029	4.002603250	5.0121	E 3.508		7.02803		E 10.65	
	H1 1/+ 99.9885*	**H2** 1+ .0115*	**H3** 1/+ 12.32 a	**H4** 2− 8E-23 s	**H5** (1/+) very short		**H6** (2 −) 3E-22 s		**6**	
	σ 332, 149	σ .52 mb, .23 mb	β− .018591 σ, < 6 μb	n	n		3n ?, 4n ?			
	1.00782503207	2.0141017778	E .018591	4.0278	5.035		6.0449			

For example, let's look at beryllium (Be). The atomic mass of ^{6}Be is 6.01973 u with its
half-life of 5.0×10^{-21} s. The black strip on ^{7}Be indicates that the isotope is naturally

available but radioactive with its half-life of 53.3 d. ^{9}Be is stable with 100% abundance and atomic mass of 9.0121821 u. Other numbers and symbols will be discussed in later.

Example 3.14

From the above figure, how many stable helium isotopes are there?

Solution

There are two stable helium isotopes; they are ^{3}He (0.000137% abundance) and ^{4}He (99.999863%).

Example 3.15

Determine the half-life of ^{7}He and its atomic mass.

Solution

The half-life of ^{7}He is 3E-21 s with its atomic mass of 7.02803 u.

Important Terms (in the Order of Appearance)

Plum Pudding	Lyman Series	Magic Numbers
Bohr's first postulate	Paschen series	Band of stability
Stationary states	Balmer series	Liquid drop model
Bohr's second postulate	Proton–electron model	Binding energy
Bohr's third postulate	Proton–neutron model	Chart of the nuclides
Bohr's radius	Nuclear force	Half-life
Balmer series	Nuclear shell model	

BIBLIOGRAPHY

Bush, H.D., *Atomic and Nuclear Physics,* Prentice Hall, Englewood Cliffs, NJ, 1962.
Ebbing, D.D., *General Chemistry 5th Ed.,* Houghton Mifflin Company, Princeton, NJ, 1996.
Mayo, R.M., *Nuclear Concepts for Engineers,* American Nuclear Society, La Grange Park, IL, 1998.
Shultis, J.K. and R.E. Faw, *Fundamentals of Nuclear Science and Engineering 2nd Ed.,* CRC Press Taylor & Francis Group, Boca Raton, FL, 2008.
Shultis, J.K. and R.E. Faw, *Radiation Shielding,* American Nuclear Society, La Grange Park, IL, 2000.
Tipler, P.A., *Physics for Scientists and Engineers,* Worth Publishers, New York, 1976.
Wapstra, A.H., "Atomic Masses of Nuclei," in *Handbuc der Physik,* S. Flugge, Ed., Vol. 38, Springer-Verlag, Berlin, 1958.

FURTHER EXERCISES

A. True or False: If the statement is false, give a counterexample or explain the correction. If the statement is true, explain why it is true.
1. The plum pudding model was developed based from results of Rutherford and Thomson's experiments.
2. Becquerel's experiment showed the discovery of beta, gamma, and alpha radiation types.
3. The gold foil experiments helped scientists to rule out the plum pudding model.
4. Light nuclei lie far away from the $N = Z$ line in the Chart of Nuclides.
5. The emitted energy of photon of the 1st line ($n = 2$ to $n = 1$) wavelength in the Lyman series is 10.2 eV.
6. ^{16}O has a magic number as both protons and neutrons indicate its stability.
7. Emission of a single proton or neutron often occurs in radioactive nuclei.
8. Heavier nuclei require an excess of protons to be stable.
9. In the "Proton–Neutron Model," a nucleus with a mass number A contains Z protons and $N = A - Z$ neutrons.
10. Generally, isotopes heavier than Pb are radioactive and will be listed with ½-life values.

B. Problems:
1. Of the following nuclides, two are radioactive. Which are radioactive and which is stable? Explain. (1) $^{118}_{50}Sn$; (2) $^{76}_{33}As$; (3) $^{227}_{89}Ac$.
2. From the given pair, $^{122}_{51}Sb$ and $^{136}_{54}Xe$, choose the nuclide that is radioactive. Explain your selection.
3. Calculate the photon energy (eV) required to excite the hydrogen electron from its ground state to the first excited orbit.
4. Calculate the photon energy (eV) for the three longest wavelengths in the Balmer series and calculate the wavelengths (nm).
5. Calculate the wavelengths (nm) of the first three spectral lines in the Lyman spectral series for hydrogen.
6. Imagine that there is a way to strip 28 electrons from copper (Cu, $Z = 29$) in a vapor of this metal. Determine the wavelengths (Å) of the spectral lines of the series for which $n = 2$, 3, and 4 at which $\Delta E = E_n - E_1$. Calculate the ionization potential (keV) for the last electron.
7. *Pair production* and *pair annihilation* are the quintessential examples of the conversion of energy into matter and vice versa. In pair production, a high energy photon's energy is converted to mass and kinetic energy of an electron–positron pair. The rest-mass energy of a positron or electron is 0.511 MeV.
   a. For pair production to occur, what must be the minimum (threshold) energy of the photon in MeV?
   b. What is the maximum wavelength in cm of a photon that can result in pair production?

c. Due to the conservation of momentum after the pair production occurs, any excess energy of the original photon results in kinetic energy that is equally distributed between the positron and electron. What would be the kinetic energy of the positron if the energy of the photon is 6.2 MeV?

8. An electron starting from rest is accelerated across a potential difference of 5 million volts. (a) What is its final kinetic energy in joules (J)? (b) What is its total energy in MeV? (c) What is its final mass in atomic mass unit (u)?

9. Circle the correct statements based on the below Chart of the Nuclides.

B7 (3/−)	B8 2+	B9 3/−	B10 3+	B11 3/−	B12 1+
3E-22 s	770 ms	8E-19 s	19.9ª	80.1ª	20.20 ms
p, α	β+ 14,···	p, 2α	σₐ 384E1, 173E1		β- 13.37,··
	(α) 8.359		σᵧ 3, 1		γ 4439,··
	(2α) 1.57		σₚ 7 mb	σᵧ 5 mb, 2 mb	(α) 2,··
			σₜ 8 mb		
7.0299	E 17.979	9.013329	10.0129370	11.0093055	E 13.369

a. ^{10}B and ^{11}B are stable isotones.
b. ^{10}B and ^{11}B are stable isotopes.
c. This element is known as boron.
d. This element is known as beryllium.
e. The half-life of ^{11}B is 80.1 years.
f. ^{9}B nuclide has 10.0129 amu.

NOTES

1 Before the establishment of the SI, the femtometer, 1 fm = 10^{-15} m, was called a "fermi" after the Italian physicist Enrico Fermi, one of the pioneers in nuclear fission reactor development.

2 For hydrogen, Z = 1, but it is convenient not to quantify Z so that the results can be applied to other hydrogen-like atoms.

4 Nuclear Energetics I—Binding Energy and Separation Energy

OBJECTIVES

After studying this chapter, the reader should be able to:

1. Gain basic knowledge on mass defect and binding energy (BE).
2. Know how to explain the significance of the average BE per nucleon plot.
3. Understand the subtle difference between BE and separation energy.

4.1 BASIC REACTIONS

There are several types of reactions. On the most basic level, there is
$$a + b + c + \ldots \rightarrow w + x + y + \ldots,$$

where "a" through "z" represent nuclides. Sub-classes of this include:

$a + b + c \ldots \rightarrow d$	Fusion
$a \rightarrow b + c + d + \ldots$	Radioactive decay
$a + b \rightarrow c + d$	Binary.

4.2 FOUR FUNDAMENTAL LAWS GOVERNING REACTIONS

There are four main conservations that we will explore throughout the chapter and next chapters. They are: (1) Conservation of Nucleons, (2) Conservation of Energy (Rest Mass Energy), (3) Conservation of Charge, and (4) Conservation of Momentum.

4.3 SCENARIO (PRELUDE)

Consider the following case. Here, the following particle and atomic masses have been measured very carefully:

Particle	Rest Mass (amu)	Rest Mass (Mev)
e-	0.000549	0.511
p	1.007276	938.3
n	1.008665	939.6
^{1}H	1.007825	938.6
^{2}H = D	2.014102	1876.1

DOI: 10.1201/9781003272588-4

We can make observations from this table. First, mass of neutron > mass of hydrogen > mass of proton. Second, mass of proton + mass of electron = 1.007825 amu. This provides a good accuracy. As we might expect, atomic hydrogen mass is almost completely accounted for by the combined mass of one proton and one electron. However, we will do the same for deuterium; that is,

$$m_p + m_e + m_n = 1.007276 + 0.000549 + 1.008665 \, amu = 2.016490 \, amu \neq m_D.$$

This is way greater than the mass of neutral atomic deuterium. This discrepancy is too large to be accounted for by an experimental uncertainty. Something important is missing!

4.4 GENERAL CONCEPT OF MASS DEFECT AND BINDING ENERGY

Generally, the sum of all constituent proton, neutron, and electron masses does not equal the atomic mass of the resultant atom:

$$Zm_p + Zm_e + Nm_n \neq m_{_Z^A X^N}.$$

Thus, rest mass is not conserved in the assembly process of the atom. This proposition must be true since, if it were not, there would be no difference between an assembled atom and the collection of sub-atomic particles. Since there is a mass reduction or **mass defect,** we know through mass/energy equivalence that there is also an energy decrement associated with the assembly of sub-atomic particles into an atom. This energy of assembly is called the **binding energy** (BE). It is a nuclear BE, to distinguish it from the atomic BE, which binds electrons to the nucleus in the atom. The nuclear BE is the energy that binds nucleons in the nucleus and would, therefore, be the energy required in order to break the nucleus of an atom into its individual nucleons.

Now, let's us move to the next section.

4.5 EINSTEIN'S THEORY ON "MASS DEFECT"

Recall Einstein's Relationship between mass and energy: $\Delta E = \Delta mc^2$. For a reaction, $\Delta m = \Delta mi - \Delta mf$ = "mass defect."
If
$\Delta m > 0$, this is known as EXOthermic (Energy is emitted).
$\Delta m < 0$, this is known as ENDOthermic (Energy is absorbed).
Here,

$$\Delta m \text{ or } \Delta = Z(m_p + m_e) + Nm_n - M_{atom} = Z \times M(^1H) + Nm_n - M_{atom}.$$

Now, we can re-visit the definition of the BE and its basic. *What's next?*

4.6 BASICS ON BINDING ENERGY

Binding energy (BE) is the energy required to disassemble a whole atom into its separate parts, typically in MeV. Here, *atomic binding energy* is equal to the energy binding the electrons to the nucleus, and it can be ignored in calculating the *nuclear binding energy* (which we shall just term the *binding energy*). It is important to note here that in calculating the nuclear BE, we must still include the rest mass of the electrons since the values in atomic mass tables (M_{atom}) include the rest mass of the entire atom (nucleus plus electrons). BE in nuclear reaction is analogous to the heat of formation in chemical reactions. It should be mentioned that chemical reactions will have mass change too small to measure (need accuracy to about 10 significant figures). Therefore,

$$BE = \Delta c^2 \left(\text{from } E = mc^2 \right)$$

$$BE = (931.5 \,\text{MeV/u}) \Delta$$

$$BE = (931.5 \,\text{MeV/u}) \left[ZM\left({}^1H_1\right) + (A - Z)m_n - M_{atom} \right]$$

This is an IMPORTANT equation and should be MEMORIZED. BE is said to put an atom in a "*negative* energy state" since external (*positive*) energy must be supplied to disassemble the atom. Moreover, $\Delta c^2/(N + Z) = $ BE per nucleon, which provides a measure of nuclear stability. The *larger* this number the *more stable* the nucleus.
  But how?

4.7 DERIVATION

To create a nucleus with Z protons,

$$Z \text{ protons} + (A - Z) \text{ neutrons} \rightarrow \text{Nucleus}\left({}^A X\right) + BE.$$

So,

$$BE = \left(\text{Proton Mass} + \text{Neutron Mass} - \text{Nucleus Mass} \right)c^2$$

$$= \left(m_p + m_n - m\left({}^A X\right) \right)c^2.$$

But there are no nuclear masses available, only *atomic masses*; thus, we have to use Conservation of Charge concept to do this calculation. Thus,

$$\frac{BE}{c^2} = Z \left[M({}_1^1 H) - m_e + \frac{BE_{1e}}{c^2} \right] + (A - Z)m_n - \left[M({}_Z^A X) - Zm_e + \frac{BE_{Ze}}{c^2} \right]$$

this can be simplified to

$$\frac{BE}{c^2} = ZM({}_1^1 H) + (A - Z)m_n - M({}_Z^A X) + \left[\frac{ZBE_{1e}}{c^2} - \frac{BE_{Ze}}{c^2} \right].$$

We can see that the difference between the BE of the electrons are so small that they can be ignored and the final expression yields

$$\frac{BE}{c^2} = ZM(^1_1H) + (A - Z)m_n - M(^A_Z X).$$

It should be noted that the atomic mass of hydrogen must be used in the calculation and the use of mass of proton is prohibited.

Now, let's do a classic example.

Example 4.1

Calculate the binding energy of $^4_2$He.

Solution

Here, we are going to write its nuclear reaction: $2p + 2n \rightarrow \,^4_2$He. So, we can express the mass defect, or BE as

$$\text{mass defect} = \frac{BE}{c^2} = 2M(^1_1H) + 2m_n - M\left(^4_2 He\right).$$

We can obtain the values from the Appendix, and the result becomes

$$2(1.007825\,u) + 2(1.008665\,u) - 4.002603\,u = 0.0303766\,u.$$

Applying the conversion – 931.5 MeV/u, the result is **28.30 MeV**. It is important to use this conversion factor instead of going through several unit conversions with mass unit and c^2.

Now, we can calculate the **binding energy per nucleon**, which is

$$\frac{BE}{\text{nucleon}} = \frac{BE}{A} = \frac{28.30}{4} = 7.075 \ \textbf{MeV/nucleon}.$$

Let's do another example.

Example 4.2

Compute the mass defect and the binding energy per nucleon for $^7_3$Li.

Solution

We can obtain the isotopic mass of ^{7}Li from the Appendix, which is 7.016004 u. Then, we will use the equation to obtain

$$\text{Mass defect} = 3(1.0078250) + 4(1.0086649) - 7.016004 = 0.04213\,u.$$

Therefore, the BE is 0.04213 × 931.5 MeV/u = **39.244 MeV** and BE per nucleon is then

$$39.244 \ \text{MeV}/7 = 5.61 \ \textbf{MeV/nucleon}.$$

Let's practice few more isotopes.

Example 4.3

Calculate the total binding energies of Ni-62, Fe-58, and Fe-56. Also, calculate the binding energy per nucleons.

Solution

Ni and Fe have 28 and 26 protons, respectively. We can calculate the binding energies for all of them.

$BE_{Ni\text{-}62} = (931.494 \text{ MeV/u}) \big[\, 28(1.007825 \text{ u}) + 34(1.008665 \text{ u}) - 61.928349 \text{ u} \,\big]$

$BE_{Ni\text{-}62} = \mathbf{545.2603 \text{ MeV}}$

$BE_{Fe\text{-}58} = (931.494 \text{ MeV/u}) \big[\, 26(1.007825 \text{ u}) + 32(1.008665 \text{ u}) - 57.933281 \text{ u} \big]$

$BE_{Fe\text{-}58} = \mathbf{509.9455 \text{ MeV}}$

$BE_{Fe\text{-}56} = (931.494 \text{ MeV/u}) \big[\, 26(1.007825 \text{ u}) + 30(1.008665 \text{ u}) - 55.934942 \text{ u} \big]$

$BE_{Fe\text{-}56} = \mathbf{492.2555 \text{ MeV}}$

<BE> = BE/(N + Z) = BE/A (where A is the atomic mass number) is the *average binding energy per nucleon*. Thus, the answers are

$$< BE_{Ni\text{-}62} > = \frac{(545.2603 \text{ MeV})}{62} = \mathbf{8.7945 \text{ MeV}}$$

$$< BE_{Fe\text{-}58} > = \frac{(509.9455 \text{ MeV})}{58} = \mathbf{8.7922 \text{ MeV}}$$

$$< BE_{FE\text{-}56} > = \frac{(492.2555 \text{ MeV})}{56} = \mathbf{8.7903 \text{ MeV}}.$$

Note: Values may vary slightly among different sources.
  What's next?

4.8 AVERAGE BINDING ENERGY PER NUCLEON VERSUS MASS NUMBER PLOT

We can calculate BE for all isotopes existed in the data library. We can plot the average BE per nucleon for each mass number. The result is shown in Figure 4.1.
  There are several key points with regards to the plot.

- BE > 0, is always true.
- BE/A is a measure of *stability* of the nucleus.
- Larger value of ratio means more stable.
- It has a broad maximum at A ~ 60 (near Cr, Mn, Fe).
- Several light nuclei have a large BE/A value, especially multiples of 4.
- For A > 60, BE/A is a weakly decreasing, smooth function of A, called **fission** zone.
- For A < 20, BE/A is steep increasing function of A with large local variation (e.g., ^{4}He), called **fusion** zone.
- Energy is released when moving from high to max or low to max; examples are:
  - Low to max: 3H + 2H → 4He + 1n ($\Delta E = 17.588$ MeV)
  - High to max: ^{235}U + ^{1}n → ^{90}Br + ^{141}La + 4^{1}n ($\Delta E \approx 200$ MeV).

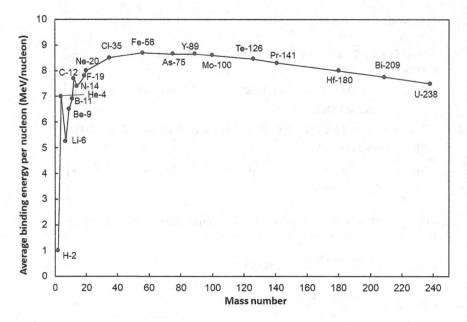

FIGURE 4.1   Avergage binding per nucleon plot.

4.9   FUSION AND FISSION

Based on the given trend in the average BE per nucleon plot, the trends can be explained qualitatively by considering the following phenomena:

- The attractive nuclear force has a very short range;
- The nuclear force is proportional to the number of nucleons; and
- The repulsive electrostatic force increases proportionally as the number of positive charges (protons) within the nucleus is increased.

From these observations, it follows that for the lightest nuclei, the combined nuclear force is limited by the small number of available nucleons. As the number of nucleons increases, this combined force grows. Since the number of protons increases as well, the repulsive electrostatic force grows simultaneously. These counteractive forces partly cancel each other and produce the broad maximum shown in the fusion zone of the plot. As the nucleus grows larger, the short-range nuclear force is increasingly unable to encompass all nucleons with equal effectiveness. We can see that the *average* BE thus starts to decrease. This implies that the nucleons, especially those that find themselves on the edges of the nucleus are held less tightly. In effect, as the nuclei increase in size beyond the broad maximum, they become less stable. These observations provide a concept that can be used in harnessing energy. First is the **fission** process where a heavy nucleus splits into two fragments, and second the **fusion** process where two light nuclei are joined to form heavier nucleus. This tends to move the elements formed toward the region of greater stability. In the end, the resultant energy release of these processes accounts for the development of nuclear power as a reality and the harnessing of fusion energy as a hope for the future.

Now, we are going to briefly discuss another concept related to BE.

4.10   NUCLEAR SEPARATION ENERGY

Closely related to the concept of BE is the **separation energy**. But what is SEPARATION ENERGY? It is the energy required to remove a single nucleon from a nucleus. Here, let us consider the addition of a single neutron to the nucleus.

$$^{A-1}_{Z}X + ^{1}_{0}n \rightarrow ^{A}_{Z}X$$

$$S_n(^{A}_{Z}X) = [m(^{A-1}_{Z}X) + m_n - m(^{A}_{Z}X)]c^2$$

$$S_n(^{A}_{Z}X) = [M(^{A-1}_{Z}X) + m_n - M(^{A}_{Z}X)]c^2$$

This can be put in terms of BE, which is

$$S_n(^{A}_{Z}X) = BE(^{A}_{Z}X) - BE(^{A-1}_{Z}X)$$

Example 4.4

What is the BE of the last neutron in a $^{16}O_8$ nucleus?

Solution:

Here, we must write the reaction, which is $^{15}_{8}O + ^{1}_{0}n \rightarrow ^{16}_{8}O$.

Then, we can use the BE equation:

$$BE = (931.5 \, \text{MeV/u})\left[M(^{15}_{8}O) + m_n - M(^{16}_{8}O) \right]$$

$$BE = (931.5 \, \text{MeV/u})[15.003065 \, u + 1.008665 \, u - 5.994915 \, u] = \textbf{15.66 MeV}$$

You can calculate BE for both ^{15}O and ^{16}O and subtract these values to get the same result as this energy is the **separation energy** since it is the energy that must be added to remove only one of the neutrons from a nuclide—in this case, the energy that must be added to remove a neutron from $^{16}_{8}O$:$^{16}_{8}O \rightarrow ^{15}_{8}O + ^{1}_{0}n$.

On another aspect, when *building* a nucleus, the individual nucleon assembly process is the minimum energy route to nucleosynthesis. Consider, for example, constructing the $^4He_2$ atom in the following three steps:

1. $^1n + ^1H \rightarrow ^2H + \gamma_1$, Q_1
2. $^1H + ^2H \rightarrow ^3He + \gamma_2$, Q_2
3. $^1n + ^3He \rightarrow ^4He + \gamma_3$, Q_3.

Note: We drop the subscript Z from the elemental symbol as that number is the same as the number of protons of that element.

The net reaction is obtained by summing these, that is,

$$2(^1n) + 2(^1H) \rightarrow ^4He + \gamma,$$

where the energy of the gamma ray γ is the sum of the excitation photon energies released in the individual synthesis steps, and

$$Q = Q_1 + Q_2 + Q_3 = BE_{4He}.$$

For this illustration, the above steps are the minimum energy path to synthesis. It can be shown that all other possible paths result in a release of greater energy.

Important Terms (in the Order of Appearance)

Mass defect	Fission
Binding energy	Fusion
Binding energy per nucleon	Separation energy

BIBLIOGRAPHY

Bush, H.D., *Atomic and Nuclear Physics*, Prentice Hall, Englewood Cliffs, NJ, 1962.

Lamarsh, J.R., and Baratta, A.J., *Introduction to Nuclear Engineering*, 3rd Ed., Prentice Hall, NJ, 2001.

Mayo, R.M., *Nuclear Concepts for Engineers*, American Nuclear Society, La Grange Park, IL, 1998.

Shultis, J.K. and R.E. Faw, *Fundamentals of Nuclear Science and Engineering 2nd Ed.*, CRC Press Taylor & Francis Group, Boca Raton, FL, 2008.

FURTHER EXERCISES

A. True or False: If the statement is false, give a counterexample or explain the correction. If the statement is true, explain why it is true.
   1. The mass of the atom is less than the sum of the masses of the protons, neutrons, and electrons.
   2. The "missing mass" of the atom is called the mass defect.
   3. The rest mass of a proton is roughly equivalent to 937 MeV.
   4. The rest masses of proton and hydrogen are identical and interchangeable.
   5. The sum of the masses of a deuterium atom's individual constituents is less than the measured mass of a neutral deuterium atom.
   6. Atomic binding energy is the energy binding the electrons to the nucleus.
   7. The energy equivalence of 1 atomic mass unit u is 931.5 MeV.
   8. Binding energy is said to put an atom in a negative energy state since external energy must be supplied to disassemble the atom.
   9. The larger the average binding energy per nucleon, the more stable the nucleus.
   10. The binding energy of Fe-58 is larger than that of Ni-62.

B. Problems:
1. Generally, energies of chemical reactions cannot be calculated by finding the difference between the masses of the reactants and the products
   because the mass must be known to 10 or more significant figures. Yet, the
   mass of the proton and hydrogen atom are known to 10 significant figures:
   Hydrogen: 1.0078250321 u, Proton: 1.0072764669 u, Electron:
   0.0005485799 u
   Estimate the binding energy of the electron BEe in the $^1_1$H atom.
2. Calculate the average binding energy (in MeV) *per nucleon* for the
   nuclides (1) $^{56}Fe_{26}$ and (2) $^{235}U_{92}$.
3. It is to our interest to conduct different experiments on $^{56}Fe_{26}$. Calculate
   the energy involved if one wants to
   a. Remove a single neutron
   b. Remove a single proton
   c. *fission* it symmetrically into two identical lighter nuclides $^{28}Al_{13}$.
      Note: The term *fission* as used in part (d) does not refer to neutron-
      induced fission (where one would expect an answer on the order
      of 200 MeV), but is used in the more generic sense of splitting or
      dividing something into two or more parts.
4. Explain why, on the curve of binding energy, the average binding
   energy per nucleon of $^4He_2$ is greater than that for $^6Li_3$? (Provide a brief
   discussion.)
5. Complete the following nuclear reactions based on the conservation of
   nucleons:
   a. $^{235}U_{92} + {}^1n_0 \rightarrow$?
   b. $? + {}^{10}B_5 \rightarrow {}^7Li_3 + {}^4He_2$
   c. $^9Be_4 + {}^4He_2 \rightarrow ? + {}^1n_0$
6. We can detect neutrons by allowing them to be captured by boron-10
   nuclei in a counter meter filled with $B^{10}F_3$ gas. The resulting nuclear
   reaction is lithium-7 and helieum-4. Compute the energy released in
   this reaction, if the relevant rest masses are: Neutron =1.00866 u, Boron-
   10 = 10.01294 u, Lithium-7 = 7.01600 u, and Helium-4 = 4.00260 u.

5 Nuclear Energetics II—Nuclear Interactions and Q-Values

OBJECTIVES

After studying this chapter, the reader should be able to:

1. Identify different types of nuclear interactions.
2. Utilize conservation of charge and nucleon to balance different nuclear interactions.
3. Develop the Q-value with the combination of conservation of mass and conservation of energy.
4. Distinguish different types of Q-value and differentiate how to calculate Q-value under different types of nuclear reactions.

5.1 COMMON USAGE

We define a nuclear *interaction* as a nuclear process that results from the collision (interaction) of two particles. In this and the remaining chapters, we often use the terms *interaction* and *reaction* interchangeably. That is why often we would see the use of nuclear reaction! In addition, the term *particle* will refer to a nucleus, nucleon, electron, positron, photon, or some other subatomic species.

5.2 NUCLEAR INTERACTIONS

We can classify the nuclear interactions as being one of the two types. First, it can be spontaneous disintegrations of nuclei, which we refer to as **radioactive decay**. In this type, the initial reactant X is a single atom that spontaneously changes by emitting one or more particles:

$$X \rightarrow Y_1 + Y_2 + Y_3 + \ldots$$

This type of reaction is exothermic in nature; that is, mass must decrease in the decay process and energy must be emitted, usually in the form of the kinetic energy of the reaction products. In general, mass of parents is greater than the sum of the product masses.

Second, it can be referred as collisions between nuclei and/or nuclear particles called interactions—absorption and scattering interactions (see Figure 5.1).

DOI: 10.1201/9781003272588-5

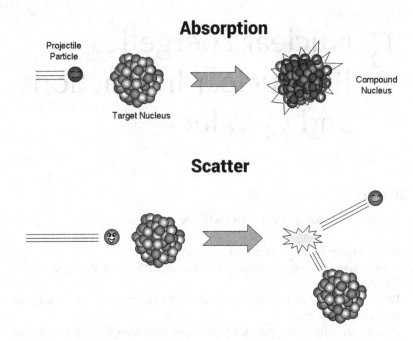

FIGURE 5.1   Two general types of nuclear collision interactions—absorption and scattering.

The general form of a collision interaction can be written as: $A + a \rightarrow (C^*) \rightarrow B + b$. A shorthand notation for this reaction is: $A(a, b)B$, where a and b are usually the lightest of the reaction pairs. Here, (C^*) is referred as a **compound nucleus** which is unstable and has a typical lifetime of $\sim 10^{-14}$ s. It should be mentioned that the process of decay of a compound nucleus is independent of mode of formation. That is, for the case of a neutron absorption, "neutrons have no memory." Thus, the above reaction mechanism is often written as

$$A + a \rightarrow B + b.$$

All these nuclear reactions follow conservation principles (including radioactive decay), which are: (1) nucleons, (2) charge, (3) total energy, and (4) linear and angular momentum.

Let's practice completing the reactions through the following the first example.

Example 5.1

Complete the following reactions:

a. $^9Be + ^4He \rightarrow ? + ^1H$
b. $^{60}Co \rightarrow ? + ^0e$
c. $^7Li + ^1H \rightarrow ? + ^4He$
d. $^{10}B + ^4He \rightarrow ? + ^1H$

Solution

General rules in balancing any nuclear reactions: (a) mass number (the number of nucleon "A") must be conserved and (b) charge (the number of proton "Z") must be conserved.

a. We have to remember that beryllium-9 has four protons, helium-4 has two protons, and hydrogen-1 has one proton. Thus, we can apply the conservation of nucleon to this reaction. So, the top LHS will be $9+4=13$ and the top RHS will be $x+1$. Thus, $x=13-1=12$. Then, we use the conservation of charge by balancing the number of protons. Therefore, $LHS=4+2=6$ and $RHS=p+1$. And $p=6-1=5$ protons. Thus, the element that has five protons is boron. The ? is ^{12}B or $^{12}_{5}B$.

b. Perform the same technique and note that cobalt has 27 protons and electron has −1 proton. Therefore, the top $LHS=60$ and the top $RHS=x+0$; $x=60$. The bottom $LHS=27$ and the bottom $RHS=y+(-1)$; $y=28$. And nickel has 28 protons; thus, $?=^{60}Ni$.

c. The answer is ^{4}He.

d. The answer is ^{13}C.

These reactions are the scattering types. So, let's look at examples of absorption reactions.

5.3   EXAMPLES OF ABSORPTION REACTIONS

Absorption reactions take the following general representation: A + a → C. We can propose the following scenario.

Example 5.2

"What target isotope must be used for forming the compound nucleus $^{60}_{28}Ni$, if the incident projectile is (a) an alpha particle, (b) a proton, and (c) a neutron?

Solution

What we are asking for is A for different cases of a to produce C. Here, C is $^{60}_{28}Ni$. Based on this information, we can perform the conservation of nucleon to show that

a. $^{4}He_{2}+^{56}Fe_{26} \rightarrow ^{60}Ni_{28}$

b. $^{1}p_{1}+^{59}Co_{27} \rightarrow ^{60}Ni_{28}$

c. $^{1}n_{0}+^{59}Ni_{28} \rightarrow ^{60}Ni_{28}$.

We can also use the basic knowledge from the Chart of the Nuclides from the previous chapter to help.

Regarding Figure 5.2, remember that the x-axis of the Chart of the Nuclides is the number of neutrons, and the y-axis the number of protons. Hence, all nuclides in a row of the chart are isotopes of the same element. For example, should the *Original Nucleus* absorb a neutron, the product nuclide is one column to the right of the

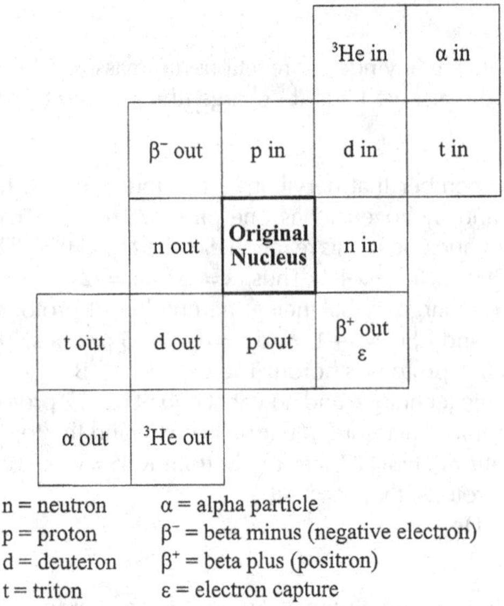

		³He in	α in
β⁻ out	p in	d in	t in
	n out	Original Nucleus	n in
t out	d out	p out	β⁺ out ε
α out	³He out		

n = neutron   α = alpha particle
p = proton   β⁻ = beta minus (negative electron)
d = deuteron   β⁺ = beta plus (positron)
t = triton   ε = electron capture

FIGURE 5.2 Relative locations of the products of various nuclear processes on the Chart of the Nuclides.

Original Nucleus, that is, the location marked "n in." Similarly, the product of a radioactive decay where the *Original Nucleus* ejects an alpha particle is the location labeled "α out." *Now, let's explore deeper into the nuclear reactions.*

5.4 MULTIPLE REACTION OUTCOMES

Some nuclear reactions can have more than a single outcome. For example, when $^{32}S_{16}$ absorbs a neutron, the possible results are as follows:

Elastic scatter: $^{1}n_0 + {}^{32}S_{16} \rightarrow {}^{32}S_{16} + {}^{1}n_0$

Inelastic scatter: $^{1}n_0 + {}^{32}S_{16} \rightarrow ({}^{32}S_{16})^* + {}^{1}n_{0'}$

($^{32}S_{16}$ is left in an excited state and neutron has less energy than if it had been elastically scattered)

(n, p) reaction: $^{1}n_0 + {}^{32}S_{16} \rightarrow {}^{32}P_{15} + {}^{1}p_1$

Radiative capture (n, γ): $^{1}n_0 + {}^{32}S_{16} \rightarrow {}^{33}S_{16} + \gamma$.

5.5 ELASTIC AND INELASTIC SCATTERING

In the multiple reactions, we can see the first two scattering types. First, we need to provide general definitions. Here, **scattering** occurs when the nucleus emits a single neutron. Typically, it is not the same neutron as the projectile neutron. The net effect is that it looks like the projectile neutron has "bounced off" the nucleus. **Elastic scatter** implies the kinetic energy of the system remains unchanged. It is similar to the classic billiard ball collision of classical physics where the total kinetic energy of the interacting particles is conserved. **Inelastic scatter** implies the kinetic energy is lost, and nucleus remains in excited state and further decays by gamma emission.

It should be noted that neutrons scattering interactions will be elastic at lower neutron kinetic energies and for lighter target nuclei, becoming inelastic at higher neutron kinetic energies and with heavier target nuclei. This is explained based on the nuclear shell model.

For light nuclei, the first excited state is typically several MeV above ground level. For example, for ^{15}O, the lowest energy level is 5.29 MeV above ground. Therefore, an interaction with a neutron whose kinetic energy is less than 5.29 MeV will be elastic. If the neutron energy is greater than 5.29 MeV, an inelastic scatter will result with the emission of a 5.29 MeV gamma ray. The energy difference between adjacent levels continues to decrease at higher energies such that it is but a few keV above 8 MeV.

Inelastic scatter is more likely in heavier elements since their energy levels are only about 0.1 MeV apart near ground level and the differences decrease to a few eV above 8 MeV.

5.6 DEFINITIONS

- **Radiative capture** is the (n, γ) reaction where the neutron is absorbed and a gamma emitted. This is a foundation of neutron activation.
- **Multiple neutron** is when 2, 3, ... neutrons are emitted from the compound nucleus. Usually endothermic, $Q < 0$, threshold energy required, KE of the system decreases (we will definitely explore the Q-value in this chapter).
- **Fission reaction**: the nucleus splits into fission fragments, neutrons and gamma radiations.

Now, let's focus on the general principles of nuclear processes.

5.7 CONSERVATION OF MASS AND ENERGY

Many of the most important laws of science are based on the "conservation" or "balance" of some property or stuff of a system as it undergoes a change. The classical laws of the conservation of energy and the conservation of mass, Bernoulli's equation, the Boltzmann particle transport equation, the Navier–Stokes equations, the conservation of momentum equation, to name but a few, are all manifestations of property conservation laws. For nuclear processes, the key law of importance is that of the "conservation of mass and energy" and the property that is being conserved is the sum of the "total" energies of the particles involved. Limiting ourselves to three types of energy,

- $m_i c^2$, rest mass energy,
- E_i, kinetic energy, and
- E^*_i, internal *excitation* energy (described further below in this section).

The law may be written as $\sum_{i=1}^{n} E_i + E_i^* + m_i c^2 = \text{constant}$ (**MI**).

We have designated the above equation as **MI**, which is an acronym for **MOST IMPORTANT!** The reader needs to understand the basis for this equation and the significance of the different terms. Here, we have ignored other forms of energy such as potential energy. Gravity, for instance, plays no significant role in nuclear reactions.

5.8   MORE ON BINARY REACTION

A most important type of nuclear reaction which is the subject of this section of course notes is the binary interaction. A **binary reaction** is one where two particles interact (i.e., A(a, b)B). The number of products of such an interaction varies depending on the specific reaction. The binary interaction is of such importance because it is by far the most *frequent* type of interaction among particles that occurs in nature. Consider the chance that three or more particles happen to come together at the same location at exactly the same time and the truth of this assertion becomes obvious.

For binary interactions, the form of conservation of mass and energy equation becomes

$$E_A + E_A^* + m_A c^2 + E_a + E_a^* + m_a c^2 = \sum_{i=1}^{n} E_i + E_i^* + m_i c^2.$$

It should be mentioned that the rest mass energy term, $m_i c^2$ for a photon (or neutrino which we typically ignore in nuclear engineering problems) will be zero. Excitation energy will be discussed later in this chapter and only applies to nuclei. For now, we can ignore it. Many binary interactions produce two products.

Case I: Consider the interaction where an oxygen-16 nucleus absorbs a neutron resulting in its transmutation to a carbon-13 nucleus and an alpha particle, that is, a helium-4 nucleus. This may be written as $^{16}O(n, \alpha)\,^{13}C$.

Notice that the total number of nucleons involved, given by the superscript atomic mass numbers (A), balance, as well as the total number of protons (or charge) involved, given by the subscript atomic numbers (Z). For this case, our general equation above reduces to

$$E_A + m_A c^2 + E_a + m_a c^2 = E_B + m_B c^2 + E_b + m_b c^2,$$

where terms with subscript A refer to the kinetic energy and rest mass energy of the oxygen-16 atom, terms with subscript a refer to the kinetic energy and rest mass energy of the incident neutron, subscript B refer to those of the carbon-13, and subscript b refer to those of the recoiling alpha. This type of nuclear interaction involving the absorption of a neutron by a nucleus can result in nuclear fission should the target nucleus be a particular type of heavy isotope. In the case of nucleus fission, more than two products result.

Consider next the case where one of the particles involved is a photon.

Case II:

$$\gamma + {}^9 Be_4 \rightarrow {}^1 n_0 + {}^8 Be_4.$$

Let the subscript A represents the photon. Since a photon has no rest mass energy, the above equation reduces to $E_A + E_a + m_a c^2 = E_B + m_B c^2 + E_b + m_b c^2$, where, given the frequency of the photon, ν, we have $E_A = h\nu$.

Case III: In the case of radioactive decay, we assume a stationary nucleus decays into two products. (There must be at least two to conserve momentum.) If subscript A represents the decaying nuclide, the above equation reduces to $m_A c^2 = E_B + m_B c^2 + E_b + m_b c^2$ since, being at rest, the decaying nucleus has no kinetic energy, that is, $E_A = 0$.

Case IV: In this final case, consider binding energy. We can think of this as a special case of a nuclear reaction where a number of individual nucleons and electrons come together to form an atom. In calculating binding energy, Eq. (MI) can be written as

$$m_A c^2 + BE_A = \sum_{i=1}^n m_i c^2,$$

where the binding energy of the resulting nucleus, BE_A, represents the mass defect which gives rise to the strong and weak nuclear forces holding the nucleus together. Since particles on the RHS of the equation are all elementary, that is, nucleons and electrons, they have no binding energy terms. Rearranging, we have the equation for binding energy.

$$BE = \sum_{i=1}^n m_i c^2 - m_A c^2.$$

What happened to the kinetic energy terms? One of the sources of kinetic energy in Eq. (MI) is the change in the total binding energy of the system of particles undergoing the nuclear reaction.

5.9   THE Q-VALUE OF A REACTION

In any reaction in an isolated system, the sum of the energy and mass energy terms of the particles involved must be conserved. The total energy *including the rest mass energy* of the initial particles must equal the total energy of the final particles. Any change to the total kinetic energy of the particles before and after the reaction must be balanced by an equivalent change in the total rest energies of the particles before and after the reaction. For now, we have ignored the possible presence of internal excitation energy.

Let E_i and $m_i c^2$ indicate the kinetic and mass-equivalent energies, respectively, of the ith participant in the reaction: $A + a \rightarrow (C)^* \rightarrow B + b$. From the conservation of mass energy, the following must be true.

$$E_A + m_A c^2 + E_a + m_a c^2 = E_B + m_B c^2 + E_b + m_b c^2$$

The **Q-value** for a nuclear reaction can be calculated either from the change in kinetic energy that occurs or the change in the total mass of the participants. In terms of a binary reaction of the form A(a, b)B, the two methods for calculating the Q-value are as follows:

$$Q = (E_B + E_b) - (E_A + E_a)$$
$$Q = (\text{kinetic energy of final particles}) - (\text{kinetic energy of initial particles}),$$
(5.1)

$$Q = [(m_A + m_a) - (m_B + m_b)] c^2$$
$$Q = (\text{rest mass of initial particles}) c^2 - (\text{rest mass of final particles}) c^2.$$
(5.2)

Alternatively, the Q-value can be calculated from the difference in the sums of the binding energies of the reactants and the products; that is,

$$Q = (BE_B + BE_b) - (BE_A + BE_a).$$
(5.3)

Equations (5.1–5.3) will provide the same results. It should be noted that Eq. (5.2) is the most common equation to be used for calculating the Q-value as one can obtain the rest mass of isotopes and/or projectile particles from the Appendix. *What's next?*

5.10   EXOTHERMIC AND ENDOTHERMIC REACTIONS

There are two ways of describing Q-value for an indication of spontaneous level of nuclear reactions.

Exothermal (or **exothermic** or **exoergic**) reaction:

$Q > 0$. In this case, the sum of the kinetic energies of the products is greater than the sum of the kinetic energies of the initial reactants.

Endothermal (or **endothermic** or **endoergic**) reaction:

$Q < 0$. These reactions have a minimum or threshold energy that must be added to the system to allow the reaction to occur.

Example 5.3

Calculate the Q-value in MeV of this reaction: $^3H + {}^2H \rightarrow {}^4He + {}^1n$. Then determine whether or not the reaction is exothermic.

Solution

Here, we will use Eq. (5.2) to calculate the Q-value. We will obtain the rest mass of each isotope from the Appendix 1B.

Hydrogen-2 = 2.014108 u; Hydrogen-3 = 3.0160493 u

Helium-4 = 4.0026032; neutron = 1.0086649 u

So, $Q = [M(^2H) + M(^3H) - M(^4He) - m_n] \times 931.5$ MeV/u = (2.014108 u + 3.0160493 u − 4.0026032 u − 1.0086649 u) × 931.5 MeV/u = (0.0188892)(931.5) = **17.595 MeV**. Since Q > 0, the reaction is **exothermic**. This reaction will occur spontaneously and energy will be released by this reaction. The complete way of writing this equation would be $^3H + ^2H \rightarrow {^4He} + {^1n} + Q$.

Example 5.4

When a 0.5 MeV neutron is being absorbed by a lithium-6 nucleus, it produces a compound nucleus that is suddenly broken up into an alpha particle without any gamma radiation.

    a. Write the nuclear reaction showing the compound nucleus as well,
    b. Calculate the Q-value and determine the energetic type of the reaction, and
    c. Write a *complete* nuclear reaction.

Solution

    a. So, we can cast the equation as

$$^1n + {^6Li} \rightarrow (C*) \rightarrow {^4He} + ?.$$

We have to use conservation of nucleon and charge to determine the unknown compound nucleus C* and ?, which can be done in two steps. First, we will determine C*. Here,

Particle/Isotope	# of Protons (Z)	Mass Number (A)	Rest Mass (u)
Neutron, n	0	1	1.008665
Lithium-6, ^{6}Li	3	6	6.01513
C*	3	7	7.01600

We can see that C* is **lithium-7**; that is, C* = **^{7}Li***.
Now, we will perform the same to determine ?; that is,

Particle/Isotope	# of Protons (Z)	Mass Number (A)	Rest Mass (u)
Neutron, n	0	1	1.008665
Lithium-6, ^{6}Li	3	6	6.01513
Helium-4, ^{4}He	2	4	4.00260
?	x	y	

We can balance the mass number, which is $1 + 6 = 4 + y$; thus, $y = 3$. We will do the same with the number of protons, which is $0 + 3 = 2 + x$; thus, $x = 1$. Therefore, ? with one proton is hydrogen and this is **hydrogen-3**, also known as tritium, ^{3}**H** (the rest mass is 3.0160493 u).

In summary, the nuclear reaction based on this experimental observation is

$$^1\text{n} + ^6\text{Li} \rightarrow ^7\text{Li*} \rightarrow ^4\text{He} + ^3\text{H}.$$

b. Now, we can calculate the Q-value of this reaction. We should simplify the equation by getting rid of the compound nucleus. The nuclear reaction that we will be determining the Q-value is 1n +^{6}Li $\rightarrow$ ^{4}He+^{3}H. We can use Eq. (5.2) to calculate the Q-value.

$$Q = \left[M\left(^6\text{Li}\right) + m_n - M\left(^4\text{He}\right) - M\left(^3\text{H}\right)\right] \times 931.5 \text{ MeV/u}$$

$$= (6.01513 \text{ u} + 1.008665 \text{ u} - 4.00260 \text{ u} - 3.01605 \text{ u}) \times 931.5 \text{ MeV/u}$$

$$= \textbf{4.80 MeV.}$$

Since $Q > 0$, this reaction is **exothermic**.

c. We can now write a complete nuclear reaction, which is

$$^1\text{n} + ^6\text{Li} \rightarrow ^7\text{Li*} \rightarrow ^4\text{He} + ^3\text{H} + Q.$$

Now, let's look into the nature of Q-value for a unique reaction, such as the radioactive decay.

5.11 Q-VALUE OF THE RADIOACTIVE DECAY

In this case, the nuclear reaction is A $\rightarrow$ B + b. Here, Eqs. (5.1) and (5.2) can be simplified to $Q = (E_B + E_b) - E_A$ and $Q = [m_A - (m_B + m_b)]c^2$, respectively. Nucleus A can generally be assumed at rest, that is, its kinetic energy is zero. That is, $E_A = 0$. Since momentum must be conserved, the vector sum of the momentums of the two decay products must equal zero, that is, they must fly off in opposite directions, and thereby, each must possess kinetic energy, so that $E_B + E_b > 0$.

Note: In the case of a moving target, that is, one with non-zero kinetic energy, the reference frame can always be transformed to one in which the target is stationary. This is because the laws of physics are invariant under any transformation to a frame moving with constant velocity, that is, an *inertial frame*. Therefore, $Q = (E_B + E_b) = [m_A - (m_B + m_b)]c^2 > 0$, and we have shown that radioactive decays are always *exothermal*.

There are several important things to consider when performing Q-value calculations.

5.12 IMPORTANT THING #1 IN Q-VALUE CALCULATION

Note that when calculating the mass defect, it is the particle *rest mass* that is specified. Hence, if one of the particles involved is a photon, the value of the mass term corresponding to the photon is zero since a photon has zero rest mass. The following is an example of this.

Example 5.5

Consider the reaction $^9Be(\gamma, n)^8Be$. Write the proper expression for the Q-value of this reaction.

Solution

Here, $Q = [(m_A + m_a) - (m_B + m_b)]c^2 = [M(^9Be) - M(^8Be) - m_n]c^2$

Assume the collision was such that the target ^9Be atom was at rest. If the resulting speeds of the ^8Be atom and neutron were v_{Be} and v_n respectively, and the wavelength of the incident gamma ray was λ; then the equation for calculating the Q-value from the particle kinetic energies would be $Q = (E_B + E_b) - (E_A + E_a)$.

$$Q = \frac{1}{2}\left[M(^8Be)v_{Be}^2\right] + \frac{1}{2}(m_n v_n^2) - \frac{hc}{\lambda}$$

Let's do this next problem.

Problem 5.1

Calculate the Q-value for $^{16}O(n, \alpha)^{13}C$. Is the reaction exothermal or endothermal? *What are your answers?*

The answers are **−2.215 MeV**, and it is **endothermal**.

How? Let's do this problem together.

Solution to Problem 5.1:

		Rest mass			Rest mass
Reactants	$^{16}O_8$	15.994915 u	Products	$^{13}C_6$	13.003354 u
	$^1n_0$	1.008664 u		$^4He_2$	4.002603 u
	Total	17.003579		Total	17.005957

$Q = (17.003579 \text{ u} - 17.005957 \text{ u}) \times 931.5 \text{ MeV/u} = \textbf{−2.215 MeV}$. This reaction is **endothermal**.

In this case, a complete reaction can be written as: $^{16}O + {}^1n + Q \rightarrow {}^{13}C + {}^4\alpha$. It is important to note that an energy of 2.215 MeV must be put into this reaction in order for it to occur.

Now, we are ready for the next important thing.

5.13   IMPORTANT THING #2

In any nuclear reaction, in which the numbers of neutrons and protons are conserved, the Q-value is calculated by replacing any charged particle rest mass by its neutral-atom counterpart rest mass. For example, proton, $m_p \rightarrow M(^1H_1)$, and alpha particle, $m_\alpha \rightarrow M(^4He_2)$. It is fortunate that Q-value can be calculated thusly since the masses of most bare nuclei are not accurately known.

To better understand why this is so, consider an example of alpha-decay, that is, radon-222 to polonium-218, which is $^{222}Rn_{86} \rightarrow\, ^{218}Po_{84} + \,^4\alpha_2$. The neutral radon has 86 electrons, neutral polonium has 84 electrons and the alpha particle, being the nucleus of a helium-4 atom, has zero electrons. In other words, according to the above equation, whereas there were 84 electrons before the radioactive decay, there are only 82 electrons remaining after the decay. What happened to the two missing electrons, or more appropriately, how do we account for the two missing electrons?

If truly an alpha particle was omitted, then the polonium-218 must be left with two extra electrons and hence have a negative charge of −2. However, we cannot look up the atomic weight of a polonium-218 atom with two extra electrons. We can look up the atomic weight of a neutral helium-4 atom. So simply transfer those two electrons to the alpha particle to convert it to a neutral helium-4 atom and the problem is solved; that is, calculate the Q-value assuming the energy–mass equivalent reaction: $^{222}Rn_{86} \rightarrow\, ^{218}Po_{84} + \,^4He_2$.

In summary, we have simply followed **Important Thing #2** by replacing the rest mass of a charged particle (the alpha particle) with the rest mass of its neutral-atom counterpart, helium-4.

Special case for changes in proton or neutron number: The above procedure (i.e., the use of neutral-atom masses) *cannot* be used for calculating Q-values when the number of protons or neutrons is *not* conserved, for example, beta decay or electron capture. These reactions involve the *weak nuclear force* and result in protons changing into neutrons or vice versa with the production of *neutrinos* or *antineutrinos*.

5.14   IMPORTANT THING #3

In reactions in which a change in proton number occurs, the conceptual addition of electrons to both sides of the reaction to form neutral atoms results in the appearance or disappearance of an extra free electron. Such reactions can be identified by the production of a neutrino or antineutrino, since the transformation of a neutron to a proton or vice versa involves a beta particle and the associated neutrino or antineutrino as follows:

$$^1p_1 \rightarrow\, ^1n_0 + \,^0e_{+1} + \nu\,(\text{neutrino})$$
$$^0n_1 \rightarrow\, ^1p_1 + \,^0e_{-1} + \nu\,(\text{antineutrino})$$

The need for the addition of one or more electrons is due to the fact that the reactions typically involve ions, that is, atoms with one or more missing or surplus electrons.

From this aspect an alpha particle can be considered a completely ionized $^4He_2$ atom and a proton an ionized $^1H_1$ atom. Atomic masses (or atomic weights) are given for neutral atoms (e.g., Table B.1 of the course text) and not ions. Hence, an electron must be added to one side or the other of the reaction equation to convert an ionized atom to a neutral atom and thereby allow the use of data from an atomic mass table.

Example 5.6

What is the Q-value of the reaction in which two protons (hydrogen nuclei) fuse to form deuteron (deuterium nucleus)?

$$^1p_1 + {}^1p_1 \rightarrow {}^2d_1 + {}^0e_{+1} + v,$$

where $^0e_{+1}$ is a positron and v is a neutrino.

Solution

Considering the protons as ionized $^1H_1$ atoms and the deuteron as an ionized deuterium $(^2H_1)$ atom, to convert them into their corresponding neutron atoms, we must add one electron to each proton and an electron to the deuteron. Since we added two electrons to the left side of the reaction, and only one to the right-side of the reaction, we need to add an additional electron to the right-side of the reaction to account for the additional electron added to the left side.

Conceptually adding two electrons to both sides of the reaction, and neglecting the electron binding energies, the equation for the Q-value calculation is now

$$^1H_1 + {}^1H_1 \rightarrow {}^2H_1 + {}^0e_{+1} + {}^0e_{-1} + v$$

$$\text{So, } Q = \left[2M\left(^1H_1\right) - M\left(^2H_1\right) - 2m_e \right]c^2,$$

where we have assumed mv (the neutrino mass) $= 0$.

$$Q = [2 \times 1.00782503\,u - 2.01410178\,u - 2 \times 0.00054858\,u](931.5\,MeV/u)$$

$$= 0.420\,MeV.$$

Now, let's look at Q-values for two last types of reactions.

5.15   Q-VALUE FOR REACTIONS PRODUCING EXCITED NUCLEI

Often a product nucleus is left in any excited state which subsequently decays from the emission of one or more γ particles returning the nucleus to its ground state.

$$^1n_0 + {}^{32}S_{16} \rightarrow \left(^{33}S_{16}\right)^* \rightarrow {}^{33}S_{16} + \gamma$$

The Q-value calculation takes account of the fact that the mass of the excited nucleus is greater than that of its corresponding ground state by an amount E^*/c^2, where E^* is the internal **excitation energy**. (This is the remaining type of energy accounted for in our general mass and energy conservation equation.)

$$M\left(^{A}X_{Z}\right)^{*}=M\left(^{A}X_{Z}\right)+\frac{E^{*}}{c^{2}},$$

where $M(^{A}X_{Z})$ is the mass of the ground state atom given in nuclide mass tables.

Example 5.7

What is the Q-value of the reaction ^{10}B(n, α)^{7}Li* in which the lithium nucleus is left in an excited state 0.48 MeV above its ground state?

Solution

We can have

$$Q=\left[m_{n}+M\left(^{10}B_{5}\right)-M\left(^{4}He_{2}\right)-M\left(^{7}Li_{3}^{*}\right)\right]c^{2}\text{ where}$$

$$M\left(^{7}Li_{3}^{*}\right)=M\left(^{7}Li_{3}\right)+(0.48\text{ MeV})/c^{2}.\text{ So,}$$

$$Q=\left[m_{n}+M\left(^{10}B_{5}\right)-M\left(^{4}He_{2}\right)-M\left(^{7}Li_{3}\right)\right]c^{2}-0.48\text{ MeV}$$

$$Q=[1.0086649+10.0129370-4.0026032-7.0160040]u\times(931.5\text{ MeV/u})$$

$$-0.48\text{ MeV}=\mathbf{2.310\text{ MeV}}.$$

And the last type is…

5.16   Q-VALUE FOR NUCLEAR FISSION

Here is a good example of a fission reaction.

$$^{235}U_{92}+{}^{1}n_{0}\rightarrow^{139}Xe_{54}+{}^{95}Sr_{38}+2\left(^{1}n_{0}\right)+7(\gamma).$$

We can calculate the Q-value equation for this reaction. That is, the prompt energy release is calculated from the mass deficit, that is, the Q-value.

$$Q=\left[M\left(^{235}U_{92}\right)+m_{n}-M\left(^{139}Xe_{54}\right)-M\left(^{95}Sr_{38}\right)-2m_{n}\right]c^{2}.$$

Inserting the atomic masses from Appendix 1.B $Q=[235.043923+1.008665-138.918787-94.919358-2(1.008665)]$ u × (931.5 MeV/u)=**183.6 MeV**, which is on the order of the 200 MeV/fission typically stated and used for a broad energy estimation.

Important Terms (in the Order of Appearance)

Radioactive decay	Multiple neutron
Compound nucleus	Fission reaction
Absorption reaction	Binary reaction
Scattering	Q-value
Elastic scatter	Exothermic
Inelastic scatter	Endothermic
Radiative capture	Excitation energy

BIBLIOGRAPHY

Bush, H.D., *Atomic and Nuclear Physics,* Prentice Hall, Englewood Cliffs, NJ, 1962.

Lamarsh, J.R., and Baratta, A.J., *Introduction to Nuclear Engineering,* 3rd Ed., Prentice Hall, NJ, 2001.

Mayo, R.M., *Nuclear Concepts for Engineers,* American Nuclear Society, La Grange Park, IL, 1998.

Shultis, J.K. and R.E. Faw, *Fundamentals of Nuclear Science and Engineering 2nd Ed.,* CRC Press Taylor & Francis Group, Boca Raton, FL, 2008.

FURTHER EXERCISES

A. True or False: If the statement is false, give a counterexample or explain the correction. If the statement is true, explain why it is true.
 1. Radioactive decay can be represented by $a \rightarrow b + c + d + \ldots$
 2. A shorthand notation of $A + a \rightarrow B + b$ is $A(a, b)B$.
 3. In every nuclear interaction, charge and mass must be conserved.
 4. When you bombarded nitrogen-14 in air with alpha particle and observed the production of proton, representing by hydrogen nuclei, you will also see oxygen-16 on the product side.
 5. Based on the given reaction, $^9Be(\alpha, n)?$, the isotope in the ? is carbon-12. The rest masses of proton and hydrogen are identical and interchangeable.
 6. An inelastic scatter implies that KE is conserved.
 7. $^{30}P \rightarrow {}^{30}Si + ?$, where? is the positron.
 8. We know that for any nuclear reaction, the total energy does not have to be conserved.
 9. In an exothermic reaction, energy converts to mass.
 10. If the reaction has $Q > 0$, then the reaction will occur spontaneously.

B. Problems:
 1. Complete the following reactions:
    a. $^4He(p, d)?$

    b. $?(\alpha, n)^{12}C$

    c. $^{14}N(n, p)?$

2. Calculate the Q-value of all three reactions in Problem 1 and identify the energetic type.

3. Look back at the nuclear reaction in Problem 1(a). Now, determine the binding energy of all isotopes and particles in the reaction. Then, use Eq. (5.3) to compute Q-value of the reaction using these results. Why do we neglect the BE of p?

4. Using $E=mc^2$, determine the energy of a chemical reaction $C+O_2 \rightarrow CO_2$ with the change of heat of reaction $\Delta H=-9.405 \times 10^4$ cal/mol. Discuss the obtain result and the conclusion you can get from this calculation.

5. Uranium-238 ($^{238}U_{92}$) is radioactive and decays by emitting the following particles in succession before reaching a stable form:

$$\alpha, \beta, \beta, \alpha, \alpha, \alpha, \alpha, \alpha, \beta, \beta, \alpha, \beta, \beta, \alpha$$

What is the final stable nucleus? Note: $\beta=e^-$.

6. Consider a section of the Chart of the Nuclides with hypothetical nuclides identified by letters as shown. In the Chart of the Nuclides, the number of neutrons increases by one with each additional column from left to right and the number of protons (i.e., the atomic number) increases by one with each additional row from bottom to top. For example, nuclides G and H are two different isotopes of the same element since they have the same number of protons but nuclide H has one more neutron than nuclide G. If nuclide M is the location of the parent nuclide, what is the letter of the location of the daughter nuclide for each of the following decay processes?

    a. Alpha particle decay

    b. Negative beta particle (electron) decay

    c. Gamma ray emission

    d. Proton emission

A	B	C	D	E
F	G	H	I	J
K	L	M	N	O
P	Q	R	S	T
U	V	W	X	Y

7. Calculate the net energy released (in MeV) for the following fusion reaction:

$$^2H + {}^2H \rightarrow {}^3He + {}^1n.$$

Is it exothermic? Explain.

8. When a projectile proton particle is interacting with beryllium-9 isotope, there are at least six possible exit channels for this reaction. That is, there are six possible reactions resulting from this initial interaction; these are:

$$^1_1p + {}^9_4Be \rightarrow \begin{array}{l} {}^{10}_5B + \gamma \\ {}^9_5B + {}^1_0n \\ {}^9_4Be + {}^1_1p \\ {}^8_4Be + {}^2_1H \\ {}^7_4Be + {}^3_1H \\ {}^6_3Li + {}^4_2He \end{array}.$$

Calculate the Q-value (in MeV) for these above reactions. Which are exothermic and which are endothermic. Only provide a detailed calculation of the first case. The remaining results may be reported in a spreadsheet table providing the atomic weights used and the resulting Q-value.

9. Calculate the Q-value (in MeV) of this beta radioactive decay:

$$^{38}Cl \rightarrow {}^{38}Ar + {}^0e + \nu.$$

Note: In beta and positron decay, the number of protons in the parent and daughter atoms is different from the number of electrons in the neutral parent and daughter atoms. The change in electron number must be accounted for. We can ignore the contribution of the neutrino and antineutrino.

10. Polonium-210 decays to the ground state of ^{206}Pb with a half-life of 138 days by the emission of a 5.305-MeV alpha particle. What is the kinetic energy of the daughter nuclide (^{206}Pb) in MeV?

6 Radioactivity and Radioactive Decay

OBJECTIVES

After studying this chapter, the reader should be able to:

1. Identify and recognize different modes of radioactive decay.
2. Know how to calculate the kinetic energy and Q-value for each decay type.
3. Explain and use energy level diagrams for various isotopes.
4. Understand the nature of radioactive decay, decay constant, half-life, and activity.
5. Understand decay chain and know how to set up a basic rate of change equation for decay dynamics.
6. Trace the decay series and use an equilibrium condition to determine concentration and activity of isotopes.
7. Identify life time of the materials using radiodating.

6.1 NUCLEAR DECAY MODES

It is important to emphasize the conservation principles of radioactive decay and nuclear reactions; these are: (1) conservation of charge; (2) conservation of mass (i.e., number of nucleons); (3) conservation of energy; and (4) conservation of linear and angular momentum.

Based on the nature of these conservation principles, of the more than 2930 known isotopes, almost 2700 are radioactive—that is, *radioisotopes*. Only 65 of known radioisotopes are naturally occurring:

- Primordial with long half-life, for example, ^{238}U and ^{40}K
- Continuously created by cosmic rays, for example, ^{3}H, ^{7}Be, and ^{14}C.

In addition, some decays can also result in internal conversion (IC), which is $\left(^{A}_{Z}P^{*} \rightarrow [^{A}_{Z}P] + e^{-}\right) ^{A}_{Z}P \rightarrow ^{A-1}_{Z}D + n$ the most common form of radioactive decay by a nuclide in an excited (metastable) state emitting a γ-ray $\left(^{A}_{Z}P^{*} \rightarrow ^{A}_{Z}P + \gamma\right)$, about 2000 examples. Importantly, from Chapter 5, we have shown that all radioactive decay modes result in the release of energy, that is, they have a Q-value > 0. For radioactive decay, it is important to know that the Q-value may be referred to as **disintegration energy** as well.

Next, we have to look back at the nuclear stability.

DOI: 10.1201/9781003272588-6

TABLE 6.1

Nuclear Decay Modes

N/Z Ratio	Decay Mode	Reaction (P = Parent Atom and D = Product or Daughter Atom)	~ Number of Examples
Neutron surplus	β^- particle (electron/negatron) $v_e^* =$ antineutrino	${}_Z^A P \rightarrow {}_{Z+1}^A D + \beta^- + v_e^*$	1100
	n (neutron)	${}_Z^A P \rightarrow {}_Z^{A+1} D + n$	~ 80
Neutron deficit	α particle (^{4}He nucleus)	${}_Z^A P \rightarrow {}_{Z-2}^{A-4} D + \alpha$	450
	β^+ particle (positron) $v_e =$ neutrino	${}_Z^A P \rightarrow {}_{Z-1}^A D + \beta^+ + v_e$	250
	Electron capture (EC)	${}_Z^A P \rightarrow {}_{Z-1}^A D^* + v_e$	
	p (proton)	${}_Z^A P \rightarrow {}_{Z-1}^{A-1} D + p$	~40

6.2 NUCLEAR STABILITY

Figure 6.1 demonstrates that with increasing atomic mass, nuclei require a greater and greater surplus of neutrons over protons to maintain stability. Radioactive decay is always such that the daughter nucleus tends to become more stable, in other words, move toward the "stable nuclide" locations. Generally, those radioactive nuclides farthest from the belt of stability have the shortest decay times. Based on this figure,

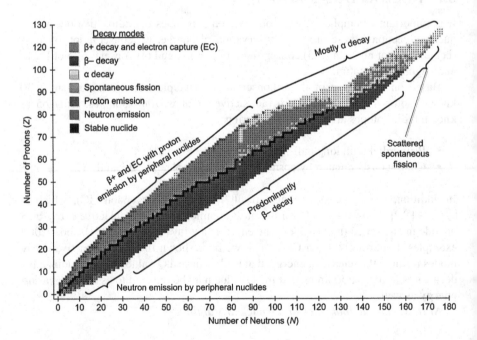

FIGURE 6.1   Nuclear stability.

it can be seen that the heavy radioactive elements and their unstable daughters emit three natural types of radiation: alpha particles, gamma rays, and beta particles.

   Let's us explore these first three natural decay!

6.3 ALPHA DECAY

Alpha decay is typical of proton-rich heavy nuclei. Initially the daughter atom still has Z electrons; that is, it is a doubly negatively charged ion. The excess electrons quickly break away from the atom leaving it in a neutral state. The doubly charged alpha particle quickly gives up its kinetic energy by ionizing and exciting atoms along its trajectory and acquires two orbital electrons to become a neutral ^{4}He atom. So, the reaction can be represented by

$$_Z^A P \rightarrow \left[_{Z-2}^{A-4} D \right]^{2-} + _2^4 \alpha \rightarrow _{Z-2}^{A-4} D + _2^4 He.$$

For example

$$_{92}^{238} U \rightarrow _{92}^{234} Th + _2^4 He$$

$$_{92}^{238} U \rightarrow _{92}^{234} Th + _2^4 \alpha$$

$$_{92}^{238} U \xrightarrow{\alpha} _{92}^{234} Th.$$

As with all types of radioactive decay, in order for alpha decay to occur, the Q-value for the radioisotope must be $Q_\alpha > 0$, that is,

$$M\left(^A P\right) > M\left(^{A-4} D\right) + M\left(^4 He\right),$$

where Q_α equals the kinetic energy of the decay products.

   Alpha particle energies are typically in the range of 10 MeV or less, so relativistic corrections are not necessary. The schematic diagram of the alpha decay is shown in Figure 6.2.

   In the following discussion, the notation for the mass has been changed from lower to upper case, for example, m_p is replaced by M_p. To simplify the calculation of the dynamics of alpha decay, we can always assume that the radioisotope undergoing the decay is at rest. Hence, its total system rest mass energy plus kinetic energy is equal to only the rest mass energy of the radioisotope, since the kinetic energy of a body at rest is zero.

   Initial total system mass plus energy = $M_p c^2$.

   Similarly, the initial total momentum of the system is zero.

   The conservation of mass energy then requires that the total system rest mass energy plus kinetic energy after the decay equals that before the decay. Since in order for the alpha particle to be emitted, it must undergo motion, it possesses both kinetic

Alpha-particle Decay

Before decay the parent nucleus is at rest.

Total system mass + energy = m_Pc^2
Total system momentum = 0

After decay the daughter nucleus and the alpha-particle are in motion.

Total system mass + energy = m_Pc^2 = $m_Dc^2 + m_\alpha c^2 + \frac{1}{2}\,m_D v_D^2 + \frac{1}{2}\,m_\alpha v_\alpha^2$
Since $Q = m_Pc^2 - m_Dc^2 - m_\alpha c^2$ then
$Q = \frac{1}{2}\,m_D v_D^2 + \frac{1}{2}\,m_\alpha v_\alpha^2$
Total system momentum = $0 = m_D v_D - m_\alpha v_\alpha$

FIGURE 6.2   Schematic summary of alpha particle decay.

energy and momentum. Hence the total system rest mass energy plus kinetic energy is given by

$$M_Pc^2 = M_Dc^2 + M_\alpha c^2 + \frac{1}{2}M_D v_D^2 + \frac{1}{2}M_\alpha v_\alpha^2 \quad \text{conservation of mass-energy.}$$

Subtracting the sum $M_Dc^2 + M_\alpha c^2$ from both sides of the equation gives

$$M_Pc^2 - M_Dc^2 + M_\alpha c^2 = \frac{1}{2}M_D v_D^2 + \frac{1}{2}M_\alpha v_\alpha^2.$$

The left-hand side of the equation is by definition the Q-value for the decay reaction.

$$Q_\alpha = \frac{1}{2}M_D v_D^2 + \frac{1}{2}M_\alpha v_\alpha^2 \tag{6.1}$$

Hence, the Q-value for the decay reaction is equal to the total kinetic energy of the system created by the loss of mass due to the radioactive decay.

Similarly, since total system momentum must be conserved (equal to zero), the momentum of the daughter isotope must equal that of the alpha particle, but be in the opposite direction.

$$0 = M_D v_D - M_\alpha v_\alpha \quad \text{conservation of linear momentum.} \tag{6.2}$$

Hence, we have two Eqs. (6.1 and 6.2) with only two unknowns, v_D and v_α, whose values can then be found. It follows that

$$Q_\alpha = \frac{1}{2} \times \frac{M_\alpha^2}{M_D} v_\alpha^2 + \frac{1}{2}M_\alpha v_\alpha^2 = \frac{1}{2}M_\alpha v_\alpha^2 \left[\frac{M_\alpha}{M_D} + 1 \right].$$

The alpha particle is emitted at a well-defined energy, that is, *the energy of the alpha particle is given by*

$$E_\alpha = Q_\alpha \left[\frac{M_D}{M_D + M_\alpha} \right] \approx Q_\alpha \left[\frac{A_D}{A_D + A_\alpha} \right].$$

Here, $E_D = Q_\alpha - E_\alpha$ and *the energy of the daughter atom is given by*

$$E_D = Q_\alpha \left[\frac{M_\alpha}{M_D + M_\alpha} \right] \approx Q_\alpha \left[\frac{A_\alpha}{A_D + A_\alpha} \right].$$

Occasionally, the alpha decay does not go directly to the ground state as indicated by the Q-value, but rather remains in an excited state for a short period which further decays by gamma emission to the ground state (e.g., energy levels for an alpha decay of ^{226}Ra). Thus, we can recast Q_α as

$$Q_\alpha = KE_D + KE_\alpha.$$

It is important to see that this $Q_\alpha = KE$ and when $\gamma = 0$, then $Q_\alpha = Q$. From here, the kinetic energy (KE) shared by the particles can be calculated because

$$Q = KE + \gamma = (m_P - m_D - m_a) \times 931.5 \, \text{MeV/u}.$$

Let's work on an example together here.

Example 6.1

^{234}U may decay by ejection of an alpha accompanied by the emission of a gamma, which in this case has an energy of 0.053 MeV. (a) Write the nuclear reaction for this situation; (b) Determine total mass energy conversion, which is Q; (c) Determine the KE shared by the particles; and (d) Determine the KE for alpha emitted by ^{234}U.

Solution

   a. We will first do the conservation of nucleon and charge to obtain a proper alpha decay reaction.
      Z: $92 = x + 2 \rightarrow x = 90$; # A: $234 = y + 4 \rightarrow y = 230$
      So, this is thorium-230 and the reaction is $^{234}_{92}\text{U} \rightarrow {}^{230}_{90}\text{Th} + {}^4_2\alpha$.
   b. We can calculate Q-value for this by obtaining the rest mass value from the Appendix 1B.

   $$Q = \left[M(U - 234) - M(Th - 230) - M(He - 4) \right] \times 931.5 \, \text{MeV/u}$$

   $$= (234.040945 \, u - 230.033126 \, u - 4.002603 \, u) \times 931.5 \, \text{MeV/u} = \textbf{4.8587 MeV}.$$

   c. Since $Q = KE + \gamma$, thus $KE = Q - \gamma = 4.8587 - 0.053 = \textbf{4.8057 MeV}$.

d. Lastly, we need to calculate KE_α, which is

$$KE_\alpha = KE\left[\frac{M_D}{M_D + M_\alpha}\right] \approx KE\left[\frac{A_D}{A_D + A_\alpha}\right] = 4.8057\left(\frac{230}{234}\right) = \mathbf{4.8057 \ MeV}.$$

Now, we are going to move to the next natural decay, a gamma type.

6.4 GAMMA DECAY

A **gamma decay** occurs when an excited nucleus decays to its ground state by the emission of a gamma photon. The lifetime of an excited state is typically within 10^{-9} sec and the reaction can be represented by

$$_Z^A P^* \text{(excited or metastable nucleus)} \rightarrow {}_Z^A P + \gamma.$$

Here, an excited state sometimes indicated by notation $(^A X)^*$ or ^{Am}X. Some excited states have lifetimes much longer than 10^{-9} sec (in the range of seconds to days). These are termed **metastable** or **isomeric** states and indicated by superscript m. For example, ^{137m}Ba or $^{60m}Co \rightarrow {}^{60}Co + \gamma$. So, we can present this LHS term by

$$M(^A P^*)c^2 = M(^A P)c^2 + E^* = M(^A P)c^2 + E_P + E_\gamma.$$

Hence, $E^* = E_P + E_\gamma$ where E_P is the recoil kinetic energy of the resulting ground state nuclide.

Since the kinematics of gamma decay are of the same form as that for alpha decay, that is, a single radioisotope decays by splitting into two entities, the mass energy and momentum conservation equations for gamma decay follow a form similar to those given above for alpha decay, the exceptions being that the gamma-ray, being a photon, has no rest mass and hence no rest mass energy, and the energy and momentum of the gamma-ray are given by $h\nu$ and $h\nu/c$, respectively, where h is Planck's constant and ν is the gamma-ray frequency.

As before, assuming that the parent nuclide is at rest,

$$M_P c^2 = M_D c^2 + \frac{1}{2} M_D v_D^2 + h\nu \quad \text{conservation of mass-energy.}$$

Subtracting the sum $M_D c^2$ from both sides of the equation gives

$$M_P c^2 - M_D c^2 = \frac{1}{2} M_D v_D^2 + h\nu.$$

The LHS of the equation is by definition the Q-value for the decay reaction.

$$Q_\gamma = \frac{1}{2} M_D v_D^2 + h\nu. \tag{6.3}$$

Similarly, from momentum considerations,

$$0 = M_D v_D^2 - \frac{hv}{c} \text{ conservation of linear momentum.} \tag{6.4}$$

Again, we have two Eqs. (6.3 and 6.4) with only two unknowns, v_D and v, whose values can then be found.

Assuming the parent nuclide is at rest, the Q-value must equal the excitation energy of the parent nuclide and in turn equal the kinetic energy produced by the decay.

$$Q_\gamma = E^* = E_P + E_\gamma$$

$$p_g = \frac{E_\gamma}{c} \text{ momentum of photon}$$

$$E_P = \frac{1}{2} M_P v_P^2 \rightarrow v_p = \left[\frac{2E_P}{M_P}\right]^{\frac{1}{2}}$$

$$M_P v_P = [2M_P E_P]^{\frac{1}{2}} \text{ momentum of recoil nucleus.}$$

Hence,

$$p = \frac{E_\gamma}{c} = [2M_P E_P]^{\frac{1}{2}} \text{ and } E_P = \frac{E_\gamma^2}{(2M_P c^2)}.$$

From above, $Q_\gamma = E_P + E_\gamma$. So,

$$E_P = \frac{E_\gamma^2}{(2M_P c^2)} = Q_g - E_\gamma$$

$$\frac{E_\gamma + E_\gamma^2}{(2M_P c^2)} = Q_\gamma \rightarrow E_\gamma \left[\frac{1 + E_\gamma}{(2M_P c^2)}\right] = Q_\gamma$$

$$E_\gamma = Q_\gamma \left[1 + \frac{E_\gamma}{2M_P c^2}\right]^{-1} \cong Q_\gamma = E^*.$$

That is, the kinetic energy of the recoil nucleus is negligible compared to the energy of the gamma-ray. In the end, $E_\gamma \cong Q_\gamma = E^*$. *Now, the third natural decay...*

6.5 BETA DECAY (ALSO KNOWN AS NEGATRON DECAY)

A **beta decay**[1] or **negatron decay** happens when a neutron in the nucleus changes to a proton and through that process an electron (β^-) and an antineutrino (v_{e*}) are emitted.

$$_Z^A P \rightarrow _{Z+1}^A D + \beta^- + v_e^*$$

We can cast this reaction in a proper way as $_Z^A P \rightarrow \left[_{Z+1}^A D \right]^+ + \beta^- + v_e^*$.

This allows us to see the balance in charge. Beta decay is typical of neutron-rich nuclides. The beta particle β^-, an electron, is apparently ejected from the nucleus, perhaps as a result of neutron decay; that is,

$$^1 n_0 \rightarrow\, ^1 p_1 +\, ^0 e_{-1} + v_e^*.$$

The fact that the decay energy is not shared in very definite proportions between the daughter nucleus and the β^- as with alpha decay implies the existence of a third particle, the antineutrino. The antineutrino v_e^* may carry off between 0 and 100% of energy (typically 2/3).

Thus, it is possible to set the KE for beta particle, which is

$$Q_\beta /c^2 = M\left(^A P_Z \right) - M\left(\left[^A D_{Z+1} \right]^+ \right) + m_\beta + m_{v*} \approx M\left(^A P_Z \right) - \left[M\left(^A D_{Z+1} \right) - m_e \right]$$
$$+ m_\beta + m_{v*}$$
$$= M\left(^A P_Z \right) - M\left(^A D_{Z+1} \right).$$

For spontaneous beta decay, $Q_\beta > 0$; that is, $M(^A P_Z) > M(^A D_{Z+1})$. Unlike for alpha and gamma decay, as demonstrated above, instead of two equations (the conservation of mass energy and momentum equations) with two unknowns, with beta decay we have two equations with three unknowns, the third unknown being the energy and (possibly) the momentum of the neutrino. Hence, the beta particle is not emitted with a definite energy in the rest frame of reference of the parent isotope, as with the alpha particle and gamma-ray, but within possible range of energies as shown in Figure 6.3.

Often in β^- decay the nucleus of the daughter is left in an excited state, for example, ^{38}Cl beta decay (see Figure 6.4). For β^- decay to an excited level with energy E^* above ground level in the daughter: $Q_\beta /c^2 = M(^A P_Z) - M(^A D_{Z+1}) - E^*/c^2$.

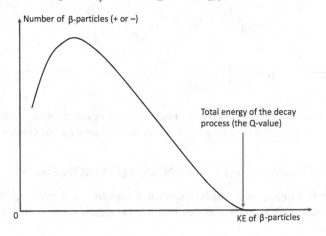

FIGURE 6.3 Distribution of the KE of beta particles.

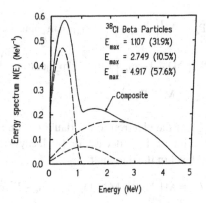

FIGURE 6.4 Energy spectra of principle chlorine-38 beta particle (Shultis and Faw, 2008).

Because the kinetic energy of the parent nucleus is assumed to be zero, Q_β decay energy must be divided among the kinetic energies of the products. The β^- particle energy is maximum when the antineutrino has negligible energy, in which case $Q = E_D + E_\beta \approx E_\beta$.

Example 6.2

Experimental observation shows that the radionuclide argon-41 decay is a beta decay type where it gets to an excited level of $^{41}K^*$ at which it is 1.293 MeV above the ground state. Write the decay and calculate the maximum KE of the emitted beta particle.

Solution

The reaction is $^{41}_{18}Ar \rightarrow \left[^{41}_{19}K \right]^+ + \beta^- + \nu^*_{e^-}$.

From the Appendix 1B, M(Ar-41) = 40.9645008 u and M(K-41) = 40.96182597 u.
Here, E* = 1.293 MeV. So,

$$Q_\beta/c^2 = M\left(^{41}Ar \right) - M\left(^{41}K \right) - E^*/c^2$$

$$Q_\beta = (40.9645008\,u - 40.96182597\,u) \times 931.5\,MeV/u - 1.293\,MeV = 1.199\,MeV.$$

Now, we can introduce a counterpart of the beta decay in the next section.

6.6 POSITRON DECAY

The reaction can be written properly as

$$^A_Z P \rightarrow \left[^A_{Z-1}D \right]^- + \beta^+ + \nu_e,$$

where ν_e is a *neutrino*. **Positron decay** is typical of proton-rich nuclides. The positron β^+ is the antiparticle of the electron. For example, $^{15}_8O \rightarrow \left[^{15}_7N \right]^- + \beta^+ + \nu_e$.

$$Q_{\beta+}/c^2 = M\left(^A P_Z\right) - M\left(\left[^A D_{Z-1}\right]^-\right) + m_{\beta+} + m\_neutrino_\beta \approx M\left(^A P_Z\right)$$

$$-\left\{\left[M\left(^A D_{Z-1}\right) - m_e\right] + m_{\beta+} + m\_neutrino_\beta\right\}$$

$$= M\left(^A P_Z\right) - M\left(^A D_{Z-1}\right) - 2m_e,$$

where the binding energy of the electron to the daughter ion and the neutrino mass have been neglected. Similar to beta decay, for β^+ decay to an excited level with energy E* above ground level in the daughter:

$$Q_{\beta+}/c^2 = M\left(^A P_Z\right) - M\left(^A D_{Z-1}\right) - 2m_e - E^*/c^2.$$

For spontaneous positron decay, $Q_{\beta+} > 0$; that is,

$$M\left(^A P_Z\right) > M\left(^A D_{Z-1}\right) + 2m_e + \frac{E^*}{c^2}.$$

The emitted positron loses its kinetic energy by ionizing and exciting atomic electrons along its trajectory. Eventually, it captures an ambient electron forming for a brief instant a pseudo-atom called *positronium* before *pair annihilation* occurs.

Now, let's explore another decay type.

6.7 ELECTRON CAPTURE (EC)

An **electron capture** (EC), also known as **K-capture**, occurs when an orbital electron is absorbed by the nucleus, converting a nuclear proton into a neutron and neutrino. Here, the innermost K-shell electrons having the greatest probability (see Figure 6.5). The capture of such an orbital electron by a proton results in

$$p + {}^0 e_{-1} \rightarrow n + v_e \text{ and } {}_Z^A P \rightarrow {}_{Z-1}^A D^* + v_e.$$

$$Q_{EC}/c^2 = M\left(^A P_Z\right) - \left\{M\left(\left[^A D_{Z-1}\right]^-\right) + m_v\right\} \approx M\left(^A P_Z\right) - M\left(^A D_{Z-1}\right)$$

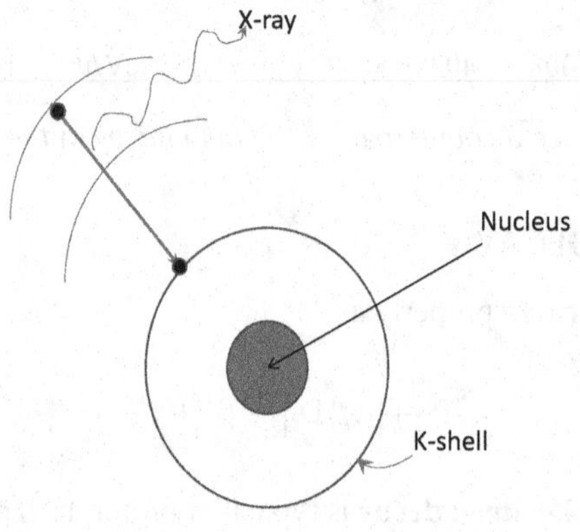

FIGURE 6.5 Mechanism of EC.

If the daughter is left in an excited state, the excitation energy E* must be included in the Q_{EC} calculation.

$$Q_{EC}/c^2 = M\left(^A P_Z\right) - M\left(^A D_{Z\text{-}1}\right) - E^*/c^2.$$

For EC to occur spontaneously, $Q_{EC} > 0$; that is, $M(^A P_Z) > M(^A D_{Z\text{-}1}) + E^*/c^2$.

Both positron decay and EC produce the same daughter nuclide. If the parent's atomic mass is not at least two electron masses greater than the daughter's mass, $Q_{\beta+}$ is negative and positron decay cannot occur. Q_{EC} is positive as long as the parent's mass is even slightly greater than the daughter's. As an example of EC, $^7Be_4 \rightarrow {}^7Li_3 + v_e$.

In EC, an orbital electron (usually from an inner shell) disappears leaving an inner electron vacancy. The remaining electrons cascade to lower orbital energy levels to fill the vacancy, usually emitting X-rays as they become more tightly bound. Instead of emitting X-rays, the energy change may also be transferred to an outer orbital electron ejecting it from the atom. The ejected electrons are called *Auger* electrons.

Example 6.3

Consider the decay of nickel-59 by EC: (a) Write the nuclear reaction; (b) Determine Q-value for this reaction; (c) Determine the recoiling velocity and kinetic energy; and (d) Show that it is safe to assume that the neutrino gets all of the kinetic energy.

Solution

a. We can write it as $^{59}_{28}Ni \rightarrow {}^{59}_{27}Co + v_e$ or $^{59}_{28}Ni + {}^{0}_{-1}e \rightarrow {}^{59}_{27}Co + v_e$.

b. Q = (58.934351 u – 58.93320) × 931.5 MeV/u = **1.0721 MeV**.

c. Here, we can us the relativistic particle of neutrino by using the momentum equation from Chapter 2: $p = \dfrac{1}{c}\sqrt{KE^2 + 2KE \times m_0c^2}$. Since the neutrino rest mass is zero, then $p_v = \dfrac{KE_v}{c}$, which may be used in the momentum balance as being equal in magnitude to the momentum of the recoiling Co nucleus:

$m_{Co}V_{Co} = KE_v/c \rightarrow KE_v = m_{Co}V_{Co}c$ (representing the neutrino energy).

The recoiling Co is nonrelativistic; thus,

$$KE_{Co} = \frac{m_{Co}V_{Co}^2}{2}.$$

The sum of the kinetic energies is equal to Q.

$$1.07 = m_{Co}V_{Co}c + m_{Co}V_{Co}^2/2 \rightarrow V_{Co}^2 + 2cV_{Co} - 2 \times 1.07/m_{Co} = 0$$

$$V_{Co}^2 + 2 \times 3 \times 10^8 V_{Co} - \frac{2 \times 1.07 \text{ MeV} \times 1.602 \times 10^{-13} \text{ kg m}^2\text{s}^{-2}\text{MeV}^{-1}}{58.9332 \text{ u} \times 1.660438 \times 10^{-27} \dfrac{\text{kg}}{\text{u}}} = 0.$$

$$V_{Co}^2 + 6 \times 10^8 V_{Co} - 3.503 \times 10^{12} = 0$$

This can be solved by using a quadratic formula to show that V_{Co} = **5.8393 × 10^3 m/s**.

The kinetic energy of the Co nucleus is calculated by

$$KE_{Co} = \frac{m_{Co}V_{Co}^2}{2} = \frac{(58.9332)(1.660438E-27)(5.8393E3)^2}{2(1.6021E-19)} = 10.41 \text{ eV}.$$

d. Compared to the total of 1,070,000 eV, this is insignificant. Initially, it would have been safe to assume that the neutrino received the total energy ($Q \approx KE$), and the momentum balance yields only the recoil nucleus velocity. Thus, it is possible to see that

$$V_{Co} = \frac{KE}{(m_{Co}c)} = \frac{(1.07 \text{ MeV})(1.6021E-13 \text{ kg m}^2\text{s}^{-2}\text{MeV}^{-1})}{\left(3E10\frac{\text{cm}}{\text{s}}\right)(58.9332 \text{ u})\left(1.660438E-27\frac{\text{kg}}{\text{u}}\right)} = 5.84 \times 10^3 \text{ m/s}.$$

So, it is safe to say that the neutrino gets all of the KE.

6.8   NEUTRON DECAY

Neutron decay occurs when neutron-rich nuclides emit a neutron usually leaving an excited daughter nucleus which subsequently emits a gamma-ray in returning to its ground state: ${}_Z^A P \rightarrow {}_Z^{A-1}D + n$. Here,

$$Q_n/c^2 = M\left({}^A P\right) - \left[M\left({}^{A-1}P*\right) + m_n\right] = M\left({}^A P\right) - M\left({}^{A-1}D\right) - m_n - E*/c^2.$$

For neutron decay to occur and leave the ground state of the daughter ($E* = 0$),

$$M\left({}^A P\right) > M\left({}^{A-1}D\right) + m_n.$$

In general, very energetic nuclei can expel a neutron (~8 MeV) and this is rarely occurred. This small but important group of neutrons simplifies the control of a reactor, for example, ${}^{138}Xe_{54} \rightarrow {}^{137}Xe_{54} + {}^1n_0$.

6.9   PROTON DECAY

This type of decay is very uncommon in nature. However, in 1970, direct proton decay was reported. Bombardment of ${}^{40}Ca$ nuclei by ${}^{16}O$ nuclei has resulted in the formation of an unstable ${}^{53}Co$ nucleus which decays to ${}^{52}Fe$ by the emission of a proton. The same nucleus has been produced by proton bombardment of ${}^{54}Fe$ nuclei to yield the same ${}^{53}Co$ nuclei and again proton decay. The half-life of the proton emitting ${}^{53}Co$ is very short (245 ms) and the decay process reduces both A and Z by one. Basically, **proton decay** happens when proton-rich nuclides emit a proton usually leaving an ionized daughter nucleus because of the extra electron. The extra electron is subsequently ejected from the atom leaving a neutral daughter: ${}_Z^A P \rightarrow \left[{}_{Z-1}^{A-1}D\right]^{-1} + p$.

$$Q_p/c^2 = M\left({}^A P\right) - \left\{\left[M\left({}^{A-1}D\right) + m_e\right] + m_p\right\} \approx M\left({}^A P\right) - M\left({}^{A-1}D\right) - M\left({}^1H\right) - E*/c^2$$

$$\approx M\left({}^A P\right) - M\left({}^{A-1}D\right) - M\left({}^1H\right) - E*/c^2.$$

For proton decay to occur to leave a ground state of the daughter ($E^* = 0$),

$$M\left(^A P\right) > M\left(^{A-1} D\right) - M\left(^1 H\right).$$

6.10 INTERNAL CONVERSION

Often the daughter is left in an excited state which decays (usually within ~10^{-9} seconds) to the ground state by the emission of one or more gamma rays. The excitation energy may also be transferred to an atomic electron (usually a K-shell electron) causing it to be ejected from the atom and leaving a singly ionized ground state. This is known as an **internal conversion** (IC). Here, $^A_Z P^* \to \left[^A_Z P\right]^+ + e^-$. Because of the relatively large binding energies for K-shell electrons for heavy nuclei BE_e, the amount of kinetic energy shared between the recoil ion and ejected electron must be accounted for.

$$\frac{Q_{IC}}{c^2} = M\left(^A_Z P^*\right) - \left\{ M\left(\left[^A_Z P\right]\right)^+ + m_e \right\} \approx \left\{ M\left(^A_Z P\right) + \frac{E^*}{c^2} \right\}$$

$$- \left[\left\{ M\left(\left[^A_Z P\right]\right) - m_e + \frac{BE_e}{c^2} \right\} + m_e \right] \approx \frac{[E^* - BE_e]}{c^2}$$

To conserve the initial linear momentum, the daughter and IC electron must divide the decay energy as

$$E_e = \left(\frac{M\left(^A_Z P\right)}{M\left(^A_Z P\right) + m_e} \right) [E^* - BE_e] \approx E^* - BE_e$$

and

$$E_D = \left(\frac{m_e}{M\left(^A_Z P\right) + m_e} \right) [E^* - BE_e] \approx 0.$$

As the outer electrons cascade down in energy to fill the inner-shell vacancy, X-rays and Auger electrons are also emitted.

6.11 ENERGY LEVEL DIAGRAM

Often in beta decay, the nucleus of the daughter is left in an excited state. As shown below for cobalt-60 (see Figure 6.6), for a decay to an energy level E^* above the ground level, the mass of the daughter atom must be replaced by the mass the *excited* daughter. It is crucial to pay attention to the energy level diagram. Figure 6.6 provides several important information about the decay of cobalt-60. In general, Q-value can be calculated to be $Q = c^2[M(\text{Co-60}) - M(\text{Ni-60})] = 931.5 \times [(59.933822) - (59.930790)] = 2.81$ MeV. We can see that from the figure, $2.505 + 0.31$ MeV ≈ 2.81 MeV as well. It is shown that the half-life of cobalt-60 is 5.272 years. The energy scale indicates that this is the beta decay (2.81 MeV) above the ground state (0 MeV) of ^{60}Ni. Arrows point to the right in this diagram since beta decay increases Z. The situation shows that the beta decay has the Q_β of 0.31 MeV and more than 99% of the time to

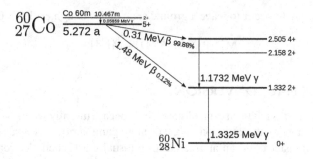

FIGURE 6.6   Decay diagram for Co-60.

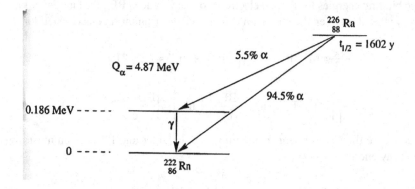

FIGURE 6.7   Decay diagram for Ra-226.

the excited state 2.5 MeV above the ground state of ^{60}Ni. The excited state immediately decays by the emission of two monoenergetic photons of 1.17 and 1.33 MeV, respectively. A tiny fraction of decay events (~0.1%) result in β^- decay directly to the state 1.33 MeV above ground, which subsequently decays by single photon emission.

Thus, the energy level diagram provides important aspects into the decay and gives us the perspective of such reaction as radioactive decay is purely statistical in nature.

Let's look at another example, Ra-226 decay's decay diagram (see Figure 6.7).

An alpha decay is shown above with the arrows point to the left since Z is reduced in a decay. We can calculate Q-value to be

$$Q = c^2 \left[M(Ra) - M(Rn) - M(a) \right] = 931.5 \text{ MeV/u}$$

$$\times \left[(226.025402 \text{ u}) - (222.017570 \text{ u}) - (4.002603 \text{ u}) \right]$$

$$= 4.870 \text{ MeV}.$$

Here, 94.5% of the ^{226}Ra decays immediately to the ground state of ^{222}Rn, and the remainder are to the excited state, 0.186 MeV above the ^{222}Rn ground state. As with excited states following beta decay, this one too will decay by γ emission. It should be noted that although ^{222}Rn is unstable and itself decays by an emission, there is no difficulty in identifying it as a ground (but unstable) state and referencing the nuclear states of ^{226}Ra to this state.

6.12 DECAY CONSTANT

The probability of radioactive decay for a given isotope has traditionally assumed to be a constant, imperious to any external influence. There is no way of predicting whether or not a single nucleus will undergo decay in a given amount of time. However, we can predict the expected behavior of a very large number of identical radionuclides. So, if $N(t)$ is the number of nuclei left at time t,

$$\frac{dN(t)}{dt} = -\lambda N(t),$$

where λ is the radioactive **decay constant** (units are 1/time → 1/s, 1/min, etc.); it is the probability per unit time that a decay will occur. If at $t = 0$, there are N_0 nuclei, then

$$N(t) = N_0 e^{-\lambda t} \text{ or } N(t) = N_0 \exp(-\lambda t).$$

There are two important points: (1) this equation is only valid for the case where the nuclide that is decaying is *not being replenished*, and (2) the same equation holds if we assume that N and N_0 represent atom number densities instead of number of atoms.

6.13 MEAN LIFETIME, HALF-LIFE, AND ACTIVITY

Example 6.4

Show that the probability that a given nucleus will decay in a time interval t to $t + dt$ is $p(t)dt = \lambda \exp(-\lambda t) dt$.

Proof: the integral of $p(t)dt$ over 0 to ∞, that is, all time, must equal 1.

$$\int_0^\infty p(t)\,dt = \lambda \int_0^\infty \exp(-\lambda t)\,dt = \lambda \times \left(\frac{-1}{\lambda}\right)\exp(-\lambda t)]_0^\infty = -\left[\frac{1}{\infty} - \frac{1}{1}\right] = 1$$

Similarly, the probability that an individual nuclide will decay between time zero and time t is

$$\int_0^\infty p(t)\,dt = \lambda \times \left(\frac{-1}{\lambda}\right)\exp(-\lambda t)]_0^t = -[\exp(-\lambda t) - 1] = 1 - e^{-\lambda t}$$

If <t> is the **mean lifetime** of a nucleus before decay, then

$$<t> = \int_0^\infty t\,p(t)\,dt = \lambda \int_0^\infty t\exp(-\lambda t)\,dt = \frac{1}{\lambda}$$

Demonstrate that

$$<t> = \lambda \int_0^\infty t\exp(-\lambda t)\,dt = \frac{1}{\lambda}$$

This is indeed the statistical average time a nucleus exists before undergoing radioactive decay. From a table of definite integrals, we find that $\int_0^\infty x^n e^{-ax}\,dx = n!/a^{n+1}$. Here, $n = 1$, $x = t$, and $a = \lambda$; therefore,

$$<t> = -\lambda \times \frac{1}{\lambda^2} = \frac{1}{\lambda}.$$

Let $T_{1/2}$ be the *radioactive* **half-life**.

$$N\left(T_{\frac{1}{2}}\right) = \frac{N_0}{2} = N_0 \exp\left(-\lambda T_{\frac{1}{2}}\right)$$

That is, the average amount of time required for a sample to decrease by 50%. Hence,

$$T_{\frac{1}{2}} = N\left(\frac{Ln2}{\lambda}\right) = \frac{0.0693}{\lambda}$$

Now, we can determine **activity**, which is the average number of disintegrations per unit time.

$$A(t) = activity = \lambda N(t) = \# \, disintegrations$$

Do not confuse activity A with gram atomic weight A. A unit of *activity* is the *curie* (Ci) where 1 Ci = 3.7 × 10¹⁰ disintegrations/s. (This is equivalent to the activity of roughly 1 g of radium.) An alternative unit is the *Becquerel* (shared 1903 Nobel Prize with the Curies for the discovery of radioactivity) where 1 Bq = 1 disintegration/s.

If *N(t)* is the number of nuclei at time t, the activity is

$$A(t) = \frac{-dN(t)}{dt} = \lambda N(t) = \lambda N_0 \exp(-\lambda t) = A_0 \exp(-\lambda t).$$

Note: Unlike with the equation for simple radioactive decay, $N(t) = N_0 \, e^{-\lambda t}$, where N can represent either the number of atoms or the atom number density; in the equation for activity, N can only represent the number of atoms.

Let's do a couple examples.

Example 6.5

$T_{1/2}(^3H_1)$ = 12.32 year (tritium). What is the activity in Ci of 1 g of tritium?

Solution

Recall Avogadro's number, N_a = 0.6022 × 10²⁴ atoms/mol
The number of atoms N(t) in a pure material of mass m is

$$N(t) = \frac{mN_a}{A} \text{ and } A(t) = \lambda N(t) = \frac{\lambda m N_a}{A}$$

$$T_{\frac{1}{2}} = \frac{(\ln 2)}{\lambda} = \frac{0.693}{\lambda}.$$

Therefore,

$$\lambda = \frac{0.693}{T_{\frac{1}{2}}} = \frac{0.693}{\left[(12.32 \text{ yr})(365.3 \text{ d/yr})(24 \text{ h/d})(3600 \text{ s/h})\right]} = 1.78 \times 10^{-9} \text{/s}$$

$$A(t) = (1.78 \times 10^{-9} \text{/s})(1 \text{ g})(0.6022 \times 10^{24} \text{ at/mol})/(3 \text{ g/mol})$$

$$= (3.58 \times 10^{14} \text{dis/s})(1 \text{ Ci}/3.7 \times 10^{10} \text{dis/s}) = \mathbf{9.67 \times 10^3 Ci}.$$

Example 6.6

a. What is the activity of 1 g of ^{238}U? (b) What is the time required for ^{238}U to decay by 1%—that is, for 1% of the atoms present at time zero to undergo radioactive decay?

Solution

a. We can approach the problem like this:

$$A(t) = \lambda N(t)$$

$$T_{\frac{1}{2}} = 4.47 \times 10^9 \, y$$

$$\lambda = \frac{0.693}{T_{\frac{1}{2}}} = 1.55 \times 10^{-10} y^{-1}$$

$$N = mN_a / A_{28} = \frac{(1\,g)(0.6022 \times 10^{24}\,at/mol)}{(238.05\,g/mol)} = 2.53 \times 10^{21} at$$

$$A(t) = 1.55 \times 10^{-10} y^{-1} \times 2.53 \times 10^{21} at \times (1\,y/\,3.15 \times 10^7 s) \times (1\,Bq/s^{-1}) = 1.24 \times 10^4 Bq$$

$$A(t) = (1.24 \times 10^4 \, Bq)(1\,Ci/3.7 \times 10^{10}\,Bq) = 3.37 \times 10^{-7} Ci$$

b. From part (a),

$$N(t) = N_0 exp(-\lambda t) \text{ where } \lambda = 1.55 \times 10^{-10} y^{-1} \left(\text{see previous example} \right)$$

$$\frac{N(t)}{N_0} = 0.99 = exp(-1.55 \times 10^{-10} y^{-1} t)$$

$$\ln(0.99) = -1.55 \times 10^{-10} y^{-1} \, t \rightarrow t = 64.8 \times 10^6 y$$

6.14   DECAY CHAIN AND NUCLIDE BALANCE EQUATION

Radioactive decay processes can be more complicated as shown in Figure 6.8.

The decaying nuclide may itself be produced by some type of source, say R(t) nuclei/cm³-s. Then the nuclide balance equation becomes

$$\frac{dN(t)}{dt} = -\lambda N(t) + R(t).$$

Consider the radioactive decay chain: X → Y → Z with decay constants λ_x, λ_y, and λ_z respectively. The appropriate equations for this chain are as follows:

$$\frac{dN_x}{dt} = -\lambda_x N_x + R_x$$

$$\frac{dN_y}{dt} = \lambda_x N_x - \lambda_y N_y + R_y$$

$$\frac{dN_z}{dt} = \lambda_y N_y - \lambda_z N_z + R_z.$$

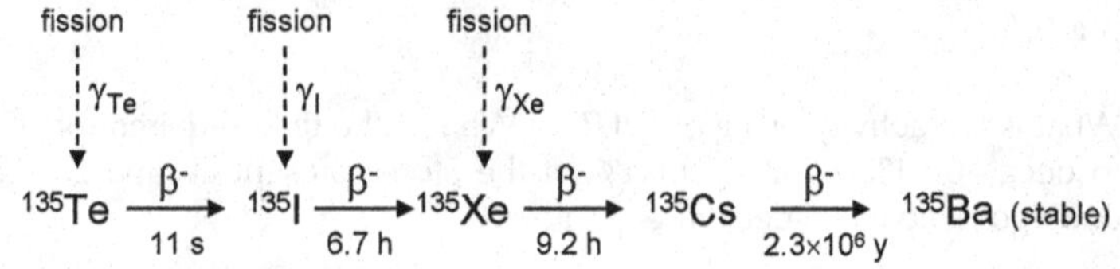

FIGURE 6.8   Example of a radioactive *decay chain* found in nuclear fission reactors.

In general, you may put them into a system of ordinary differential equations. For 2–3 chain reactions, it is convenient to solve them by hand as will be shown in the following example. However, for more chain reactions (>4), programming is necessary and solving may be done by any numerical method (e.g., Runge–Kutta).

Example 6.7

The decay chain for radioisotopes X, Y, and Z is X → Y → Z, where there are no source terms (i.e., $R_x = R_y = R_z = 0$), and Z is a stable end product.
    The set of equations describing this chain is as follows:

$$dN_x/dt = -\lambda_x N_x \tag{6.5}$$

$$dN_y/dt = \lambda_x N_x - \lambda_y N_y \tag{6.6}$$

$$dN_z/dt = \lambda_y N_y. \tag{6.7}$$

Find the expression of N_x, N_y, and N_z as the function of time.

Solution

From above, we know that the solution to Eq. (6.5) is

$$N_x(t) = N_{x0} \exp(-\lambda_x t). \tag{6.8}$$

Substituting for N_x from Eq. (6.8) into Eq. (6.6) gives

$$dN_y/dt = \lambda_x N_{x0} \exp(-\lambda_x t) - \lambda_y N_y \text{ or } dN_y/dt + \lambda_y N_y = \lambda_x N_{x0} \exp(-\lambda_x t). \tag{6.9}$$

Multiplying Eq. (6.9) through by $\exp(\lambda_y t)$ gives

$$\exp(\lambda_y t)dN_y/dt + \lambda_y N_y \exp(\lambda_y t) = \lambda_x N_{x0} \exp(\lambda_y - \lambda_x)t \text{ or}$$

$$\frac{d}{dt}[N_y \exp(\lambda_y t)] = \lambda_x N_{x0} \exp(\lambda_y - \lambda_x)t.$$

This can be integrated directly to give

$$N_y \exp(\lambda_y t) = [\lambda_x/(\lambda_y - \lambda_x)]N_{x0} \exp(\lambda_y - \lambda_x)t + C,$$

where C is an integrating constant. Multiplying through by $\exp(-\lambda_y t)$ gives

$$N_y(t) = [\lambda_x/(\lambda_y - \lambda_x)]N_{x0}\exp(-\lambda_x t) + C\exp(-\lambda_y t). \quad (6.10)$$

The value for C is found by noting that $N_y(t = 0) = N_{y0}$. Therefore,

$$C = N_{y0} - [\lambda_x/(\lambda_y - \lambda_x)]N_{x0}.$$

Inserting this into Eq. (6.10) and rearranging gives the value of N_y as a function of time.

$$N_y(t) = [\lambda_x/(\lambda_y - \lambda_x)]N_{x0}[\exp(-\lambda_x t) - \exp(-\lambda_y t)] + N_{y0}\exp(-\lambda_y t) \quad (6.11)$$

The number of atoms of the third nuclide, $N_z(t)$, is found by inserting Eq. (6.11) for N_y into Eq. (6.7) and integrating. This gives

$$N_z = \{[\lambda_x/(\lambda_y - \lambda_x)]N_{x0} - N_{y0}\}\exp(-\lambda_y t) - [\lambda_y/(\lambda_y - \lambda_x)]N_{x0}\exp(-\lambda_x t) + D, \quad (6.12)$$

where D is an integration constant determined by the conditions $N_z(t=0) = N_{z0}$. It turns out that $D = N_{z0} + N_{y0} + N_{x0}$. Inserting this expression for D into Eq. (6.12) and doing a lot of algebra finally gives

$$N_z = N_{z0} + N_{y0}[1 - \exp(-\lambda_y t)]$$
$$+N_{x0}\{1 + [\lambda_x/(\lambda_y - \lambda_x)]\exp(-\lambda_y t) - [\lambda_y/(\lambda_y - \lambda_x)]\exp(-\lambda_x t)\}. \quad (6.13)$$

Equations (6.8), (6.11), and (6.13) represent the problem solution.

Now, consider the special case where only atoms of nuclide X are present initially. Hence, N_{y0} and N_{z0} are both zero, and Eqs. (6.11) and (6.13) reduce to

$$N_y(t) = [\lambda_x/(\lambda_y - \lambda_x)]N_{x0}[\exp(-\lambda_x t) - \exp(-\lambda_y t)] \quad (6.14)$$

and

$$N_z = N_{x0}\{1 + [\lambda_x/(\lambda_y - \lambda_x)]\exp(-\lambda_y t) - [\lambda_y/(\lambda_y - \lambda_x)]\exp(-\lambda_x t)\}. \quad (6.15)$$

Let's try another example!

Example 6.8

^{135}Xe can be produced either directly as a fission product or by beta decay of ^{135}I, as indicated by the decay chain below:

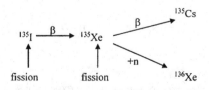

Assume that R_I represents the rate of production by fission of ^{135}I, R_{Xe} the rate of production by fission of ^{135}Xe, and L_{Xe} the loss rate of ^{135}Xe due to neutron absorption. Write the *rate* (i.e., differential) equations describing the concentrations of ^{135}I and ^{135}Xe based on this decay scheme.

Solution

For ^{131}I, the rate equation is $dN_I/dt = -\lambda_I N_I + R_I$. And for ^{135}Xe,

$$dN_{Xe}/dt = \lambda_I N_I - \lambda_{Xe} N_{Xe} + R_{Xe} - L_{Xe}.$$

Here, the change in the ^{135}Xe concentration is equal to the gain in atoms from the decay of ^{135}I, minus the loss of atoms from radioactive decay to ^{135}Cs, plus the gain in atoms being produced as a fission product, minus the loss in atoms due to neutron absorption (i.e., transmutation to ^{136}Xe). *Now, let's learn a special activity case in the next* section.

6.15 SATURATION ACTIVITY

Consider a radioactive isotope (e.g., a fission fragment) being produced at some constant rate A_0.

$$dN(t)/dt = A_0 - \lambda N(t)$$

Multiplying both sides by the integrating factor $\exp(\lambda t)$,

$$\frac{d\left[N(t)e^{\lambda t}\right]}{dt} = \left[\frac{dN(t)}{dt} + \lambda N(t)\right]e^{\lambda t} \rightarrow \frac{d\left[N(t)e^{\lambda t}\right]}{dt} = A_0 e^{\lambda t}.$$

Applying the boundary condition, $N(0) = 0$, then $\lambda N(t) = A_0[1 - \exp(-\lambda t)]$. As time approaches infinity ($t \rightarrow \infty$), $\lambda N(\infty) = A_0$. This is known as **saturation activity**.

Figure 6.9 provides an illustration of saturation activity of iodine-131.

6.16 SECULAR EQUILIBRIUM

Example 6.9

Consider a radioactive decay chain where a *daughter decays much more rapidly than the parent*. Show that the activity of the daughter approaches that of the parent, that is, that the condition of **secular equilibrium** exists.

Proof: Here, we have a radioactive decay chain such that X → Y → Z → and so forth.

Assuming no external source of the daughter, nuclide Y, other than the buildup from the radioactive decay of the parent nuclide X, the decay equation for the daughter nuclide is

$$dN_y(t)/dt = \lambda_x N_x(t) - \lambda_y N_y(t).$$

The solution is

$$N_y(t) = [\lambda_x/(\lambda_y - \lambda_x)]N_{x0}[\exp(-\lambda_x t) - \exp(-\lambda_y t)] + N_{y0}\exp(-\lambda_y t).$$

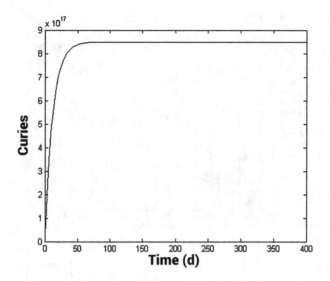

FIGURE 6.9 Saturation activity of iodine-131. ^{131}I ($T_{1/2} = 8.05\,d$) is being produced in a reactor at a constant rate of 8.5×10^{17} nuclei/s.

By definition, the activity of nuclide Y at time t is given by

$$A_y(t) = \lambda_y N_y(t) = [\lambda_y/(\lambda_y - \lambda_x)]A_{x0}[\exp(-\lambda_x t) - \exp(-\lambda_y t)] + A_{y0}\exp(-\lambda_y t),$$

where $A_{x0} = \lambda_x N_{x0}$ and $A_{y0} = \lambda_y N_{yo}$. (Note the switch in the prefix numerator term from λ_x to $\lambda_{y!}$)

The decay rate of the daughter depends on how much larger or smaller its half-life is to that of the parent. Rewrite the above equation for activity as

$$A_y(t) = A_{y0}\exp(-\lambda_y t) + [\lambda_y/(\lambda_y - \lambda_x)]A_{x0}\exp(-\lambda_x t)\,[1 - \exp\text{-}(\lambda_y - \lambda_x)t)].$$

Under the assumption that the daughter decays more rapidly than the parent, we have $(\lambda_y - \lambda_x) > 0$. As t becomes large, $A_y(t) \rightarrow [\lambda_y/(\lambda_y - \lambda_x)]A_{x0}\exp(-\lambda_x t) = [\lambda_y/(\lambda_y - \lambda_x)]A_x(t)$.

That is, the daughter decays asymptotically at almost the same rate as the parent. In the extreme case that the daughter decays *much more rapidly* than the parent, i.e., $\lambda_y \gg \lambda_x$ implies that $[\lambda_y/(\lambda_y - \lambda_x)] \approx 1$. Therefore,

$$\mathbf{A_y(t) \sim A_x(t)}.$$

The activity of the daughter approaches that of the parent; in other words, **secular equilibrium** has been established.

Aside: *secular equilibrium* can be used to determine decay constants for long-lived nuclides; for example, $\lambda_{238}N(U)_{238} = \lambda_{226}N(Ra)_{226}$; ^{238}U $T_{1/2} = 4.46 \times 10^9$ y; ^{226}Ra $T_{1/2} = 1600$ y.

These two figures (Figure 6.10) show radioactive decay series (full and partial) for uranium-238. In the long decay chain for a naturally radioactive element such as ^{238}U, where all the elements in the chain are in secular equilibrium, each of the generations

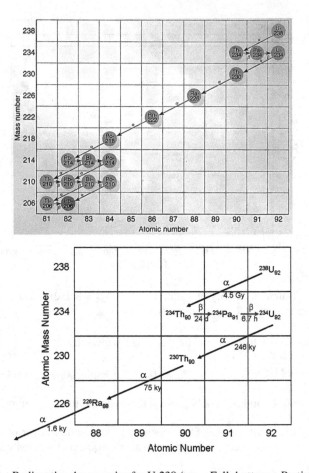

FIGURE 6.10 Radioactive decay series for U-238 (top—Full; bottom—Partial).

has built up to an equilibrium amount and all decay at the rate set by the original parent. The only exception is the final stable n^{th} on the end of the chain. Its number of atoms is constantly increasing. In such a chain, it can be represented as

$$\lambda_1 N_1 = \lambda_2 N_2 = \lambda_3 N_3 = \cdots = \lambda_{n-1} N_{n-1} \text{ or } A_0 = A_1 = A_2 = \cdots = A_{n-1}.$$

6.17   RADIODATING

An application of radioactive decay is in the dating of geological and archaeological specimens. In general, if a sample contains two radioisotopes of different decay constants, neither of which is being replenished, then we can write an equation relating the age of the sample t to the relative abundance of the two radioisotopes. That is,

$$\frac{N_1(t)}{N_2(t)} = \frac{N_{01}e^{-\lambda_1 t}}{N_{02}e^{-\lambda_2 t}}.$$

6.18 ^{14}C DATING

It is an excellent radionuclide for dating the more recently deceased. Cosmic rays continually bombard the earth with neutrons. The neutrons interact with atmospheric nitrogen to produce ^{14}C whose half-life is 5730 years. The ^{14}C interacts with oxygen to form CO_2 which is absorbed by organic matters (plants/biomass). $N_{14}/N_C = 1.23 \times 10^{-12}$ (roughly 1 ^{14}C atom for every 10^{12} atoms of ^{12}C). When the organism dies, the ^{14}C concentration diminishes through nuclear decay. It is convenient to measure the *specific activity* of ^{14}C in a sample (A_{14} per gram of carbon). Thus,

$$\frac{A_{14}}{g(C)} = \left(\frac{N_{14}}{N_C}\right)\frac{\lambda_{14}N_a}{12}$$

$$= \frac{\left(1.23 \times 10^{-12}\right)\left(\frac{0.693}{5730}\right)\left(6.022 \times 10^{23}\right)}{12} = 7.465 \times 10^6 \frac{dis}{y \times g(C)} \times \frac{1y}{3.1557 \times 10^7 s}$$

$$= 0.236 \frac{dis}{s \times g(C)} \text{ or } \frac{Bq}{g(C)} = 6.39 \times 10^{-12} \frac{Ci}{g(C)} \text{ or } 6.39 \frac{pCi}{g(C)}.$$

And the age of the artifact can be obtained by the following formula:

$$t = -\frac{1}{\lambda} \ln \frac{A_{14}(t)/g(C)}{6.4}.$$

Here, $A_{14}(t)/g(C)$ is today's specific activity.

Example 6.10

Recently a piece of old wood was excavated in the central US and found to have ^{14}C activity of 0.971 pCi/g of carbon. Estimate the age of this wood.

Solution

We note that the specific activity is 0.971×10^{-12} Ci/g(C). So, we can use the above equation: $t = -\frac{1}{\lambda} \ln \frac{A_{14}(t)/g(C)}{6.4}$, which is $t = -\frac{5730}{0.693} \ln \left(\frac{0.971}{6.4}\right) = \textbf{15,591 years}$.

   This should be considered as an uncorrected "radiocarbon age." In fact, ^{14}C production in the atmosphere has varied because of fluctuations in solar activity and in the earth's magnetic field. Recent combustion and previous nuclear weapon tests have added artificially to atmosphere concentrations.

Important Terms (in the Order of Appearance)

Disintegration energy	Proton decay
Alpha decay	Internal conversion
Gamma decay	Decay constant
Metastable or isomeric	Mean lifetime
Beta decay or negatron decay	Half-life
Positron decay	Activity
Electron capture or K-capture	Saturation activity
Neutron decay	Secular equilibrium

BIBLIOGRAPHY

Bush, H.D., *Atomic and Nuclear Physics*, Prentice Hall, Englewood Cliffs, NJ, 1962.

Coombe, R.A., *An Introduction to Radioactivity for Engineers*, Macmillan Publishing Co., Inc., New York, 1968.

Lamarsh, J.R., and Baratta, A.J., *Introduction to Nuclear Engineering, 3rd Ed.*, Prentice Hall, Hoboken, NJ, 2001.

Libby, W.F., *Radiocarbon Dating, 2nd Ed.* University of Chicago Press, Chicago, IL, 1955.

Mayo, R.M., *Nuclear Concepts for Engineers*, American Nuclear Society, La Grange Park, IL, 1998.

Shultis, J.K. and R.E. Faw, *Fundamentals of Nuclear Science and Engineering, 2nd Ed.*, CRC Press Taylor & Francis Group, Boca Raton, FL, 2008.

FURTHER EXERCISES

A. True or False: If the statement is false, give a counterexample or explain the correction. If the statement is true, explain why it is true.
1. Radioactivity is due to the decay of unstable nuclei.
2. Electron capture occurs when the excitation energy of a nucleus is used to eject an orbital electron (usually a K-shell) electron.
3. During the negatron decay, neutrino will be emitted.
4. Three common emissions of radiation occurring naturally are gamma-ray, alpha particles, and beta particles.
5. An isomeric transition is associated with a gamma decay.
6. Uranium-234 is going through an alpha decay where the daughter product is thorium-230.
7. Radioactive decay is not statistical in nature.
8. If the half-life of a certain isotope is 4 days, all isotope will be gone after 16 days.
9. Secular equilibrium occurs when the parent has an extremely long half-life and the activity of the parent is larger than that of the daughter.
10. Activity can be expressed in term of Curie and Becquerel.

B. Problems
   1. Write the reaction base on the decay type for the following cases:
      a. Sodium-22 with positron decay
      b. Chlorine-38 with beta decay
      c. Radium-224 with alpha decay
      d. Argon-41 with beta decay
   2. Calculate the Q-value (in MeV) for decay reactions in Problem 1(a) and (b).
   3. The radioisotope ^{224}Ra decays by α emission primarily to the ground
      state of ^{220}Rn (94% probability) and to the first excited state of 0.241
      MeV above the ground state (5.5% probability). What are the energies of
      the two associated α particles in MeV? The radioactive decay diagram
      for the problem is:

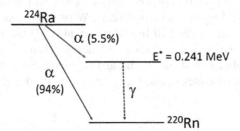

   4. The radioisotope ^{41}Ar decays by β^- (negative beta) emission to an
      excited level of 41K that is 1.293 MeV above ground state. What is the
      maximum kinetic energy of the emitted β^- particle? The radioactive
      decay diagram for the problem is:

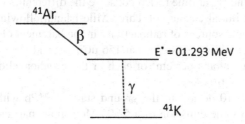

   5. The activity of a radioisotope is found to decrease by 30% in one week.
      What are the values of its (a) decay constant, (b) half-life, and (c) mean
      life? Express all answers in units of seconds or reciprocal seconds.
   6. How many grams of ^{32}P are there in a 5 mCi source?
   7. A 6.2 mg sample of ^{90}Sr (half-life 29.12 y) is in secular equilibrium with
      its daughter ^{90}Y (half-life 64.0 h). (a) How many Bq of ^{90}Sr are present?
      (b) How many Bq of ^{90}Y are present? (c) What is the mass of ^{90}Y pres-
      ent? (d) What will the activity of ^{90}Y be after 100 y?
   8. Since the half-life of ^{235}U (7.13 × 10^8 years) is less than that of ^{238}U (4.51
      × 10^9 years), the isotopic abundance of ^{235}U in natural uranium has been

steadily decreasing since the earth was formed. How many years ago was the isotopic abundance of ^{235}U in natural uranium equal to 3.0 atom percent, that is, the enrichment of uranium used in many nuclear power plants? (Assume that uranium is composed only of ^{235}U and ^{235}U isotopes.)

9. Consider a radioactive series such that isotope A decays into isotope B which in turn decays into isotope C. The first two members of the series, A and B, have half-lives of 5 hours and 12 hours, respectively, while the third member is stable. After hour many hours will the number of atoms of the second member (B) reach its maximum value if at time zero the sample size of isotope B is zero?

10. Charcoal found in a deep layer of sediment in a cave is found to have an atomic ^{14}C/^{12}C ratio only 30% that of a charcoal sample from a higher level with a known age of 1850 years. What is the (radiocarbon) age of the deeper layer? The half-life of carbon-14 is 5730 years.

11. Consider the following radioactive decay chain of fission products, where nuclide D is stable. A_0, B_0, and D_0 are the constant fission rates of production (nuclides/s), and λ_A, λ_B and λ_C are the decay constants.

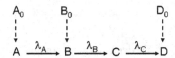

12. Write the set of differential equations describing the radioactive decay chain in terms of the above parameters, and the number of nuclei N_A, N_B, N_C, and N_D at time t. (Do not solve the differential equations.)

13. After the initial cleanup of Three-Mile Island following the accident, the principle sources of radioactivity in the basement of the unit 2 containment building were ^{137}Cs at 156 μCi/cm^3 and ^{134}Cs at 26 μCi/cm^3. How many atoms per cm^3 of each of these radionuclides were in the water at that time?

14. Polonium-210 decays to the ground state of ^{206}Pb with a half-life of 138 days by the emission of a 5.305-MeV alpha particle. A sample of ^{210}Po produces 1 MW of thermal energy from its radioactive decay. (a) What is the activity of the polonium sample in decays per second? (b) What is the mass of the ^{210}Po sample? (c) What is the kinetic energy of the daughter nuclide (^{206}Pb) in MeV?

NOTE

1 This decay process may be referred to as *electron decay, beta-minus decay, negatron decay, negative electron decay, negative beta decay*, or *simply beta decay*.

7 Binary Nuclear Reactions

OBJECTIVES

After studying this chapter, the reader should be able to:

1. Further understand the role of compound nucleus and its functions and derive expressions of energy exchange within binary nuclear reactions.
2. Understand the importance of kinematic threshold energy and Coulombic threshold energy.
3. Explain and understand various effects of neutron fissions using energy level diagrams for various isotopes.
4. Learn about the average energy loss and importance of lethargy.
5. Gain further understanding on nuclear fission and its fundamentals.
6. Understand basic fusion reactions and confinement types.

7.1 THE COMPOUND EFFECT

We need to begin again by discussing about **conservation quantities** in binary reactions involving only the *nuclear force*; these are: (1) total energy (rest mass energy plus kinetic and potential energies), (2) linear momentum, (3) angular momentum (spin), (4) charge, (5) number of protons*, (6) number of neutrons*, and (7) quantum mechanical wave function parity*. The * implies that it is not necessarily conserved in reactions involving the *weak nuclear force* responsible for beta and electron-capture radioactive decays.

Next, it is important to understand the underlying reaction mechanisms. Incident nucleons or light nuclei with kinetic energies that are *greater* than about 40 MeV have de Broglie wavelengths λ comparable to the size of nucleons in the target nucleus; hence, such an interaction involves just one or at most a few nucleons. Such reactions are considered to be **peripheral processes**. And incident nucleons or light nuclei with kinetic energies that are *less* than a few MeV have de Broglie wavelengths λ much greater than the size of nucleons in the target nucleus; hence, they interact with the nucleus as a whole producing a **compound nucleus**.

7.2 FURTHER DEPTH INTO THE COMPOUND NUCLEUS

We introduced compound nucleus in Chapter 4. We will go deeper with other terms associating with **compound nucleus**. First, we know that it decays within $\sim 10^{-14}$ seconds of formation. This is related with nuclear lifetime, which is the time it would take for an incident particle to pass through a target nucleus on the order of 10^{-21} to 10^{-17} seconds. Since this is significantly shorter than the observed lifetime of a compound nucleus, this provides credible evidence that a compound nucleus can indeed

DOI: 10.1201/9781003272588-7

be created as a result of a binary collision. Next, we have to understand about **virtual states**, which are excited states of a compound nucleus as compared to **bound states** of excited nuclei, which decay only by γ emission. The observation that the lifetime of a compound nucleus is significantly greater than the *nuclear lifetime* implies that the nucleus has <u>no</u> memory of how it was formed or incoming direction of the incident particle. Lastly, various decays from the compound nucleus can be described by **exit channels**—refer to the different modes by which a compound nucleus decays. For example, we can see three exit channels for the compound nucleus $^{64}Zn^*$.

$$
\left.\begin{array}{c} p + ^{63}Cu \\ \alpha + ^{60}Ni \end{array}\right\} \longrightarrow {}^{64}Zn^* \longrightarrow \left\{\begin{array}{l} ^{63}Zn + n \\ ^{62}Cu + n + p \\ ^{62}Zn + 2n \end{array}\right.
$$

7.3 THE BEGINNING OF KINEMATICS OF BINARY TWO-PRODUCT NUCLEAR REACTIONS

Consider the binary reaction,

$$a + A \rightarrow b + B \qquad \text{shorthand notation: A(a, b)B.}$$

From above, applying the conservation of mass and energy, this can be written as

$$E_A + M_A c^2 + E_a + m_a c^2 = E_B + M_B c^2 + E_b + m_b c^2$$

where E is kinetic energy and M or m is rest mass. If the number of protons is conserved (true except for weak force interactions), the nuclear rest masses may be replaced by the corresponding atomic masses M_i or m_i (electron masses cancel and small differences in electron binding energies are negligible):

$$Q = [(M_A + m_a) - (M_B + m_b)]c^2$$

Assume that the target nucleus is stationary in laboratory (lab) frame of reference; that is, $E_A = 0$. Thus, for the *elastic scatter*, $Q = 0$, $m_a = m_b$, and $M_A = M_B$. For the *inelastic scatter*, $Q < 0$, $m_a = m_b$, and $M_A < M_B$ since B is an excited configuration of target A.

7.4 DERIVATION OF A GENERAL EQUATION DESCRIBING THE ENERGY EXCHANGE IN A BINARY SCATTERING REACTION

Consider a *non-relativistic* binary reaction with the target nucleus at *rest* ($E_A = 0$), where p is momentum (see Figures 7.1 and 7.2).

$$E_a = E_b + E_B - Q \qquad (7.1)$$

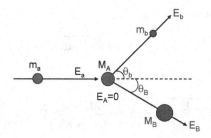

FIGURE 7.1 Diagram of a binary reaction with two products, assuming the target nucleus is at rest.

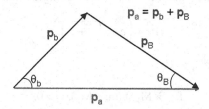

FIGURE 7.2 Momentum conservation diagram for the above reaction.

$$p_a = p_b + p_B$$

$$p_B^2 = p_a^2 + p_b^2 - 2p_a p_b \cos(\theta_b)$$

Since the reaction is non-relativistic, $p_i = [2m_i E_i]^{1/2}$.

$$2M_B E_B = 2m_a E_a + 2m_b E_b - 2[4m_a m_b E_a E_b]^{1/2} \cos(\theta_b)$$

Substituting for E_B from Eq. (7.1)

$$(M_B + m_b)E_b - 2[m_a m_b E_a]^{1/2}\omega_b[E_b]^{1/2} + E_a(m_a - M_B) - M_B Q = 0,$$

where $\omega_b = \cos(\theta_b)$. This is a quadratic equation for $\sqrt{E_b}$ where the solution must be real and positive to be physically realistic.

$$\sqrt{E_b} = \sqrt{\frac{m_a m_b E_a}{(m_b + M_B)^2}}\,\omega_b$$

$$\pm \sqrt{\frac{m_a m_b E_a}{(m_b + M_B)^2}\omega_b^2 + \left[\frac{M_B - m_a}{(m_b + M_B)}E_a + \frac{M_B Q}{(m_b + M_B)}\right]}$$

(7.2)

The solution has the form:

$$\sqrt{E_b} = c \pm \sqrt{c^2 + d}, \tag{7.2*}$$

where

$$c = \sqrt{\frac{m_a m_b E_a}{(m_b + M_B)^2}} \, \omega_b \text{ and } d = \frac{M_B - m_a}{(m_b + M_B)} E_a + \frac{M_B Q}{(m_b + M_B)}.$$

7.5   EXOERGIC AND ENDOERGIC REACTIONS

Exoergic reaction happens when $Q > 0$ and $M_B > m_a$. Here, the d term is always positive and only the positive sign of the $\pm$ pair is meaningful; that is,

$$E_b = \left[c + \sqrt{c^2 + d} \right].$$

As bombarding energy $E_a \rightarrow 0$,

$$E_b \rightarrow M_B Q/(m_b + M_B), \; Q > 0.$$

This equation represents a model of the radioactive decay case wherein the radioisotope is stationary and its mass is simply equal $(m_a + M_A)$. See, for example, the case for alpha-particle decay in the section on radioactivity. Now, for the case of zero linear momentum, $Q = E_b + E_B$ and $\theta_b + \theta_B = \pi$. E_b is the same for all angles θ_b.

   Endoergic reaction occurs when $Q < 0$. If E_a is too small, no reaction is possible. We can see that, in order for the reaction to occur in this case, the incident projectile must supply a minimum about of kinetic energy and this reaction threshold will be discussed next.

7.6   THRESHOLD ENERGY

When $Q \geq 0$ (exoergic), *except* when $M_B < m_a$, the reaction may proceed at any reactant energies, providing the Coulomb barrier condition is satisfied. For $Q < 0$ (endoergic), (1) no reaction is possible if E_a is too small; (2) energy conservation alone dictates that E_a must be greater than $-Q$; (3) this is a necessary, but *not* sufficient condition for the reaction to proceed; and (4) momentum must also be conserved—that is, Eq. (7.2) must be satisfied.

   So, it is important to introduce the **kinematic threshold energy**—for $Q < 0$ and $Q > 0$ when $M_B < m_a$. In general, the threshold energy is the minimum incident particle kinetic energy required to cause the reaction.

$$E_a^{th} = \frac{m_b + M_B}{m_b + M_B - m_a} Q \tag{7.3}$$

Since $m_i \gg Q/c^2$ for most nuclear reactions, $m_b + M_B \approx m_a + M_A$ and $m_b + M_B - m_a \approx M_A$.

Thus, Eq. (7.3) simplifies to

$$E_a^{th} \approx -\left(1 + \frac{m_a}{M_A}\right)Q. \quad (7.4)$$

7.7   DERIVATION OF EQ. (7.3) FROM (7.2)

Example 7.1

Show that starting with the general kinematic equation for a binary collision, Eq. (7.2), the minimum incident particle energy (kinematic threshold energy) for the case where $Q < 0$ or $Q > 0$ when $M_B < m_a$ is given by Eq. (7.3).

Proof: Starting with Eq. (7.2), we will express it in the form of Eq. (7.2*). For an endoergic reaction, Q is a negative value. Hence, the term under the radical in Eq. (7.2*), $c^2 + d$, will be negative if E_a is too small, so no reaction is possible. This will also be true for exoergic reactions ($Q > 0$) if $M_B < m_a$ since the first term of d will then be negative.

Hence, we set the expression under the radical equal to or greater than zero.

$$\frac{m_a m_b E_a}{(m_b + M_B)^2}\omega_b^2 + \frac{M_B - m_a}{(m_b + M_B)}E_a + \frac{M_B Q}{(m_b + M_B)} \geq 0$$

Substituting $\cos\theta$ for ω_b and multiplying both sides by $(m_b + M_B)^2$ gives

$$m_a m_b E_a \cos^2\theta + (M_B - m_a)(m_b + M_B)E_a + M_B Q(m_b + M_B) \geq 0.$$

Moving the rightmost term to the right side of the equation and factoring out E_a from the remaining left side terms gives

$$E_a[m_a m_b \cos^2\theta + (M_B - m_a)(m_b + M_B)] \geq -M_B Q(m_b + M_B).$$

Now we can use the fact that $(MB - m_a)(m_b + M_B) = M_B m_b + M_B^2 - m_a m_b - m_a M_B$.

Substituting for $(M_B - m_a)(m_b + M_B)$ in the equation gives

$$E_a[m_a m_b \cos^2\theta + M_B m_b + M_B^2 - m_a m_b - m_a M_B] \geq -M_B Q(m_b + M_B).$$

Rearranging the terms inside the brackets [...] and dividing both sides by M_B gives

$$E_a[(m_a m_b/M_B)\cos^2\theta - (m_a m_b/M_B) + m_b + M_B - m_a] \geq -M_B Q(m_b + M_B).$$

From trigonometric identities, $\cos^2\theta - 1 = -\sin^2\theta$. Therefore,

$$(m_a m_b/M_B)\cos^2\theta - (m_a m_b/M_B) = -(m_a m_b/M_B)\sin^2\theta.$$

Substituting the sine term into the above equation gives

$$E_a[-(m_a m_b/M_B)\sin^2\theta + m_b + M_B - m_a] \geq -(m_b + M_B)Q.$$

The limiting condition is when the angle of scatter of m_b, θ, is zero, thereby making the two sides of the inequality equal. Since the $\sin^2(0)=0$, the equation now becomes

$$E_a[m_b + M_B - m_a] = -(m_b + M_B)Q.$$

Solving for E_a, the kinematic threshold energy, completes the proof.

$$E_a^{th} = -\frac{m_b + M_B}{m_b + M_B - m_a}Q \qquad (7.3)$$

Alternatively, we could start earlier with the assertion that the angle is zero for the limiting condition; that is, under either of these two conditions, the minimum value of E_a (the kinematic threshold energy) for which the term under the radical in Eq. (7.2*), c^2+d will not be negative, will occur when $c^2+d=0$ and the angle of scatter of m_b, θ_b, is zero. The latter condition corresponds to $\omega^2_b=1$, since $\omega_b=\cos(\theta_b)$. Otherwise, $\omega^2_b=<1$ which corresponds to a condition where E_a is less than the kinematic threshold energy. Hence, setting $c^2+d=0$ with $\omega^2_b=1$ yields the equation:

$$\frac{m_a m_b}{(m_b + M_B)^2}E_a + \frac{M_B - m_a}{(m_b + M_B)}E_a + \frac{M_B Q}{(m_b + M_B)} = 0.$$

Multiplying both sides by $(m_b + M_B)^2$,

$$m_a m_b E_a + (m_b + M_B)[(M_B - m_a)E_a + M_B Q] = 0$$

$$E_a[m_a m_b + m_b M_B - m_a m_b + M_B^2 - m_a M_B] = -M_B(m_b + M_B)Q.$$

Dividing both sides by M_B,

$$E_a[m_b + M_B - m_a] = -(m_b + M_B)Q \rightarrow E_a^{th}$$

$$= -\frac{m_b + M_B}{m_b + M_B - m_a}Q. \qquad (7.3)$$

7.8 COULOMB BARRIER THRESHOLD

If the incident particle is a nucleus (positive charge $Z_a e$), it must overcome the Coulomb repulsive force of target nucleus of charge $Z_A e$ given by

$$F_C = \frac{Z_a Z_A e^2}{4\pi\omega_0 r^2}.$$

Work done by the incident particle starting infinitely far from target against the target's electric field at distance d is

$$W_C = -\int_\infty^d F_C \times dr = -\frac{Z_a Z_A e^2}{4\pi\omega_0}\int_\infty^d \frac{dr}{r^2} = \frac{Z_a Z_A e^2}{4\pi\omega_0 d}.$$

Hence, the kinetic energy of the incident particle is reduced by an amount W_C. A necessary condition for a reaction to occur is that the incident particle must possess kinetic energy equal to W_C sufficient to get close enough to the nucleus (distance d) such that the strong nuclear force becomes dominant.

To a first approximation, assume that $d = R_a + R_A$ where R_a and R_A are the radii of the incident and target particles, respectively. From the approximation equation for nuclei radii,

$$d = R_a + R_A = R_0 \left(A_a^{\frac{1}{3}} + A_A^{\frac{1}{3}} \right)$$

$$E_a^C \cong W_c = 1.2 \frac{Z_a Z_A}{A_a^{\frac{1}{3}} + A_A^{\frac{1}{3}}} (\text{MeV}). \tag{7.5}$$

The energy expended to penetrate the Coulomb barrier is transferred to the target nucleus which recoils and gains kinetic energy lost by the incident particle. The momentum of the incident particle equals that of the compound nucleus (subscript cn):

$$m_a v_a = M_{cn} V_{cn}$$

The kinetic energy of the compound nucleus is $E_{cn} = E_a^C (m_a / M_{cn})$. The remainder of E_a^C becomes the excitation energy of the compound nucleus. Upon decay, E_a^C is recovered in the mass and kinetic energy of the reaction products. In general,

$$\left(E_a^{th} \right)_{min} = \max \left(E_a^C, E_a^{th} \right).$$

Example 7.2

Consider the binary reaction $^{14}C(p, n)^{14}N$, where p indicates a proton and n a neutron. The neutron emerges at an angle of 90° with respect to the direction of travel of the incident proton. (a) Determine for incident proton kinetic energies of 2.8 and 0.5 MeV whether the reaction is possible; (b) If the reaction is possible, what are the kinetic energies in MeV of the resultant neutron and ^{14}N nucleus? (c) If the reaction is possible, what is the angle of the trajectory of the ^{14}N nucleus with respect to the direction of travel of the incident proton?

Solution

a. Label the reactants and products as: $^{14}C(p, n)^{14}N \rightarrow A(a, b)B$.

$Q = (m_a + M_A - m_b - M_B)$ (931.5 MeV/u)

$Q = (1.007825\ u + 14.003242\ u - 1.008665\ u - 14.003074\ u)(931.5\ \text{MeV/u})$

$= -0.626\ \text{MeV}$

The kinematic threshold energy is

$$E_{th} \approx -(1 + m_a/m_A)Q = -(1 + 1.01/14.0)\,(-0.626\text{ MeV}) = 0.671\text{ MeV}.$$

The Coulomb threshold energy is

$$E^C = \frac{(1.2)(1)(6)}{\left[1^{\frac{1}{3}} + 14^{\frac{1}{3}}\right]}\text{ MeV} = \frac{7.2}{[1 + 2.41]}\text{ MeV} = 2.11\text{ MeV}$$

$$\left(E_a^{th}\right)_{min} = \max\left(E_a^C, E_a^{th}\right) = \max\left(0.671\text{ MeV}, 2.11\text{ MeV}\right) = 2.11\text{ MeV}.$$

Hence, a reaction with a 0.5 MeV proton **is not possible** as the kinetic energy of the incident proton is less than the reaction's threshold energy of 2.11 MeV.

b. For the 2.8 MeV incident proton case,

$$\sqrt{E_b} = \sqrt{\frac{m_a m_b E_a}{(m_b + M_B)^2}}\,\omega_b \pm \sqrt{\frac{m_a m_b E_a}{(m_b + M_B)^2}\,\omega_b^2 + \left[\frac{M_B - m_a}{(m_b + M_B)}E_a + \frac{M_B Q}{(m_b + M_B)}\right]}$$

or $E_b = \left[c + \sqrt{c^2 + d}\right]^2$,

where $c = \sqrt{\dfrac{m_a m_b E_a}{(m_b + M_B)^2}}\,\omega_b$, $d = \dfrac{M_B - m_a}{(m_b + M_B)}E_a + \dfrac{M_B Q}{(m_b + M_B)}$, ω_b equals the cosine of the neutron angle of scatter, 90°, E_b is the energy of the neutron, and E_a the energy of the incident proton. Here, $\omega_b = 0$ since the cosine of 90° is zero; therefore, the any term involving with ω_b is zero. The equation reduces to

$$E_b = d = \frac{M_B - m_a}{(m_b + M_B)}E_a + \frac{M_B Q}{(m_b + M_B)} = 1.84\text{ MeV}.$$

Since $Q = E_{out} - E_{in}$,

$$-0.626\text{ MeV} = E14 + 1.84\text{ MeV} - 2.8\text{ MeV}$$

$$E_{14} = 2.8\text{ MeV} - 1.84\text{ MeV} - 0.626\text{ MeV} = \textbf{0.334 MeV.}$$

c. To calculate the scattering angle ϕ of the N^{14} nucleus, we note the following momentum balance:

$$\mathbf{p}_p = \cos(\phi)\mathbf{p}_N. \tag{7.6}$$

That is, the momentum of the incident proton must equal the momentum component of the ^{14}N nucleus in the positive x-direction. As a check we note that, since the angle of scatter of the neutron was 90°, the momentum of the resultant neutron must equal the momentum component of the ^{14}N nucleus in the positive y-direction.

$$p_n = \sin(\phi) p_N$$

In terms of the kinetic energy and rest mass, the magnitude of the momentum p is

$$p = (2mE)^{1/2}. \qquad (7.7)$$

Substituting for the momentum expression from Eq. (7.7) into Eq. (7.6) gives

$$(2m_p E)^{1/2} = \cos(\phi)(2m_N E_{14})^{1/2}$$

$$\phi = \arccos[(1.01 \times 2.8)/(14.0 \times 0.334)]^{1/2} = \arccos(0.778) = \mathbf{38.9°}.$$

Let's explore another example.

Example 7.3

Among the reactants producing the compound nucleus ^{15}N* are ^{13}C(d, t)^{12}C, ^{14}C(p, n)^{14}N, and ^{14}N(n, α)^{11}B. What is the minimum kinetic energy of the incident deuteron, proton, or neutron for each of these reactions to occur?

Solution

We calculate the Q-value for each reaction, and the kinematic threshold and Coulomb threshold energies. The minimum kinetic energy of the reaction products is

$$\min(E_b + E_B) = Q + \max(E_a^C, E_a^{th}).$$

Reaction Path	Q-Value (MeV)	E_a^C (Mev)	E_a^{th} (MeV)	Reaction Condition	$\min(E_b + E_B)$ (MeV)
^{13}C(d, t)^{12}C	1.311	1.994	0	$E_a > E_a^C$	$1.311 + 1.994 = \mathbf{3.305}$
^{14}C(p, n)^{14}N	-0.6259	2.111	0.6706	$E_a > E_a^C$	$2.111 - 0.6259 = \mathbf{1.485}$
^{14}N(n, α)^{11}B	-0.1582	0	0.1695	$E_a > E_a^{th}$	$0.1695 - 0.1582 = \mathbf{0.0113}$

It is important to note that the problem assumes that the incident particle's kinetic energy is the *minimum* necessary to allow each of the reactions to occur. This condition thereby results in the kinetic energy of the products being the *minimum* allowable. For an exothermic case with a neutral incident particle, this corresponds to the incident particle having *zero* kinetic energy ($E_a = 0$), so the total kinetic energy of the products in this case would simply equal Q. Also note that Coulomb threshold energy only exists for the case of a *positively charged incident particle*. If the incident particle were negatively charged, for example, a free electron, it would experience an increasing force of attraction toward the positively charged target nucleus as the incident particle approached it.

7.9 THRESHOLD ENERGY REVIEW

- Neutral particles—no Coulomb barrier:
  - $Q > 0$ and $M_B > m_a$, no threshold energy for the incident particle
  - $Q < 0$ or $Q > 0$ with $M_B < m_a$, kinematic threshold energy
- Positively charged particles:
  - Coulomb barrier must be overcome
  - $Q > 0$ and $M_B > m_a$, no kinematic threshold energy
  - $Q < 0$ or $Q > 0$ with $M_B < m_a$, both kinematic threshold, E_a^{th}, and Coulomb threshold, E_a^C
  - Minimum threshold $= \max(E_a^{th}, E_a^C)$

Kinematic threshold

$$E_a^{th} = -\frac{m_b + M_B}{m_b + M_B - m_a} Q \qquad (7.3)$$

$$\text{If } m_i \gg \frac{Q}{c^2}, \; E_a^{th} \approx = \left(1 + \frac{m_a}{M_A}\right) Q. \qquad (7.4)$$

Important Note: Eqs. (7.3) and (7.4) are applicable only for endoergic reactions—that is, $Q < 0$. They are *not* valid for the case of $Q > 0$ with $M_B < m_a$ (please see Figure 7.3).

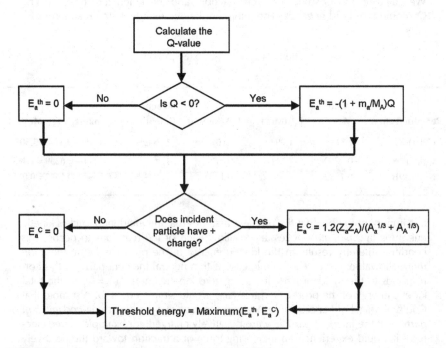

FIGURE 7.3 Summary of threshold energy calculations.

Coulomb threshold

$$E_a^C \cong W_c = 1.2\frac{Z_a Z_A}{A_a^{\frac{1}{3}} + A_A^{\frac{1}{3}}}(MeV) \qquad (7.5)$$

7.10 HEAVY PARTICLES SCATTERING (SLOWING DOWN) FROM AN ELECTRON

A heavy charged particle (e.g., an alpha particle) interacts through the Coulomb force with the electrons of the atoms in the material it is passing through. For heavy charged particles with energies in the MeV range, the atomic electrons can be considered as "free" electrons at rest due to their much smaller binding energy. Thus, $Q=0$ for the resultant scattering action. In this scenario, it can be shown that the maximum energy of the recoil electron is

$$(Ec)_{max} = 4m_e E_M/M,$$

where E_M and M are the kinetic energy and mass of the incident particle, respectively. Let's illustrate this concept.

Example 7.4

What is the maximum energy loss for a 4-MeV alpha particle scattering from an electron?

Solution

First, we use the fact that $(Ec)_{max} = 4m_e E_M/M$. Thus,

$$(Ec)_{max} = 4(0.0005486\ u)(4\ MeV)/(4.003\ u) = \mathbf{0.0022\ MeV}.$$

Since most collisions transfer from the alpha particle even less energy than this, we see that typically tens of thousands of ionization and excitation interactions are necessary for a heavy charged particle with several MeV of kinetic energy to slow down and become part of the ambient medium.

7.11 NEUTRON INTERACTIONS

As will be described in the next frame on interactions of radiation with matter, an important type of neutron interaction in the majority of existing nuclear reactor designs is elastic scatter. This is because the probability of a neutron producing a fission reaction in isotopes such as uranium-235 is hundreds of times greater for a low-energy neutron (referred to as a **thermal neutron**) than for neutrons at the energies at which they are born from the fission process. This process whereby high-energy neutrons consistently lose energy through elastic scatter collisions is termed **neutron moderation** or **neutron thermalization**. An important design element of reactors that take advantage of the high fission probability of thermal neutrons is allowing for the efficient moderation of neutrons.

7.12   NEUTRON SCATTERING AND THE WAY TO SET UP EQ. (7.2)

The two types of scattering collisions, not only for neutrons but for most types of particles, are elastic scatter and inelastic scatter; we described their basics in the previous chapter. Now, we are going to explore a bit more in kinematic aspects of these scattering types.

Elastic Scattering is referred to as an (n, n) type of the binary reaction, which is $^1n_0 + {}^AX_Z \to {}^1n_0 + {}^AX_Z$. Here, the kinetic energy of the system is *conserved* ($Q = 0$).

Inelastic Scattering is referred to as an (n, n') type, which is $^1n_0 + {}^AX_Z \to {}^1n_0 + ({}^AX_Z)^*$. The kinetic energy of the system is reduced by the excitation energy and $({}^AX_Z)^* \to {}^AX_Z + \gamma$, where $({}^AX_Z)^*$ denotes an *excited state*.

We can use Eq. (7.2) to neutron scattering, where $m_a = m_b = m_n =$ neutron mass, $\theta_b = \theta_s =$ neutron scattering angle, $E_a = E =$ neutron energies before scattering, and $E_b = E' =$ neutron energies after scattering. For elastic scatter, $M_A = M_B = M$. For inelastic scatter, $M_A \approx M_B = M$ since $Q < 0$; but we have to keep in mind that $Mc^2 \ggg |Q|$. Now, if one let $A = M/m_n \approx$ atomic mass number of the scattering nucleus, then Eq. (7.2) may be written as follows:

$$E' = \frac{1}{(A+1)^2}\left\{\sqrt{E}\cos\theta_s \pm \sqrt{E(A^2 - 1 + \cos^2\theta_s) + A(A+1)Q}\right\}^2. \qquad (7.8)$$

It should be noted that the right-hand side term inside the braces {} is squared.

7.13   ELASTIC SCATTERING OF NEUTRONS

For the case of elastic scatter ($Q = 0$), and only the plus (+) sign in Eq. (7.8) is physically meaningful. The *minimum* neutron energy (or conversely, the maximum neutron energy loss) resulting from an elastic scatter occurs when the neutron rebounds directly backward,

$$\theta_s = \pi \ (\text{i.e.,} \cos\theta_s = -1).$$

For $Q = 0$ and retaining only the plus (+) sign, Eq. (7.8) becomes

$$E' = \frac{1}{(A+1)^2}\left\{\sqrt{E}\cos\theta_s + \sqrt{E(A^2 - 1 + \cos^2\theta_s)}\right\}^2.$$

Substituting $\cos\theta_s = -1$,

$$E' = \frac{1}{(A+1)^2}\left\{-\sqrt{E} + \sqrt{E(A^2 - 1 + 1)}\right\}^2$$

$$= \frac{1}{(A+1)^2}\left\{-\sqrt{E} + \sqrt{EA^2}\right\}^2 = \frac{\sqrt{E}^2}{(A+1)^2}\left\{\sqrt{A^2} - 1\right\}^2,$$

which can be simplified to

$$E' = E(A-1)^2/(A+1)^2.$$

Hence, the *minimum* neutron energy resulting from an elastic scatter is $E' = E'_{min} = \alpha E$, where $\alpha = (A-1)^2/(A+1)^2$. The *maximum* possible neutron energy (or conversely the minimum neutron energy loss) from an elastic scatter is for a glancing interaction, $\theta_s = 0$ (i.e., $\cos\theta_s = 1$). Substituting $\cos\theta_s = 1$ into E' yields

$$E' = \frac{1}{(A+1)^2}\left\{\sqrt{E} + \sqrt{E(A^2-1+1)}\right\}^2$$

$$= \frac{1}{(A+1)^2}\left\{\sqrt{E} + \sqrt{EA^2}\right\}^2 = \frac{\sqrt{E}^2}{(A+1)^2}\left\{\sqrt{A^2}+1\right\}^2.$$

Hence, the *maximum* possible neutron energy resulting from an elastic scatter is $E = E'_{max} = E$; that is, the neutron loses no energy.

For the special case of hydrogen ($A = 1$), we find that the allowed angle of scatter is between $0 \le \theta_s \le \pi/2$. So just as in the game of billiards, there can be no backscatter (see Figure 7.4). So, $E'_{min} = 0$ and $E'_{max} = E$.

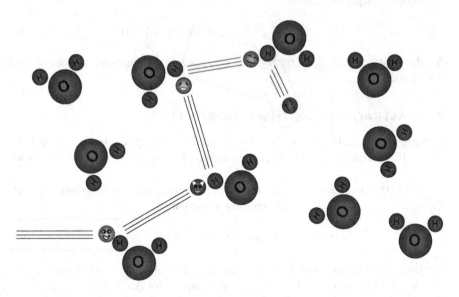

FIGURE 7.4 Illustration of energy loss of a neutron through repeated elastic scattering from the nuclei of the hydrogen atoms of water.

Example 7.5

A 4-MeV neutron is scattered elastically by ^{16}O through an angle of 90 degrees. What is the scattered neutron's energy?

Solution

Using Eq. (7.8) from above,

$$E' = \frac{1}{(A+1)^2}\left\{\sqrt{E}\cos\theta_s \pm \sqrt{E\left(A^2-1+\cos^2\theta_s\right)+A(A+1)Q}\right\}^2.$$

$Q=0$, only the plus (+) sign is retained, and $\theta_s=90°$. Therefore, $\cos(\theta_s)=0$. Here, $E=4.0$ MeV and $A=16$. Since it is *elastic* scatter, $Q=0$.

$$E' = \frac{1}{(16+1)^2}\left\{\sqrt{4\text{MeV}}\,(0)+\sqrt{(4\text{MeV})\left(16^2-1+0\right)+16(16+1)(0)}\right\}^2$$

$$= \frac{1}{289}\left\{0+\sqrt{1020}\right\}^2$$

$$E' = \left(3.46\times10^{-3}\right)(1020\text{ MeV}) = \textbf{3.53 MeV}.$$

7.14   AVERAGE NEUTRON ENERGY LOSS

From above for *elastic* scatter, $E'_{min}=\alpha E$ and $E'_{max}=E$. For **isotropic scatter** (i.e., a scattering collision in which there is equal probability of scattering through any angle), the average neutron energy loss (equal to the average kinetic energy of the recoiling target nucleus) is

$$(\Delta E)_{av} = E - E'_{av} = E - \tfrac{1}{2}(E+\alpha E) = \tfrac{1}{2}(1-\alpha)E.$$

As the *neutron energy decreases, the average energy loss due to scatter also decreases.*

7.15   AVERAGE LOGARITHMIC ENERGY LOSS

Lethargy, $u=\ln(E/E')$, is also called the logarithmic energy decrement. Here, $u=0$ for $E=E'$, and increases with decreasing energy, E. From this expression, $u=\infty$ for $E=0$.

We use the Greek letter Xi, ξ, to represent an average increase in lethargy per collision, sometimes termed the **slowing-down decrement**.

$$\xi = <\ln(E/E')>$$

For isotropic scattering with respect to the target nucleus, $\xi=1+[\alpha/(1-\alpha)]\ln\alpha$, where $\alpha=(A-1)^2/(A+1)^2$. Here, $\xi=1$ for $A=1$ (hydrogen) and for $A>1$, it is

$$\xi \approx 2/(A+2/3).$$

Several observations can be seen from this expression: (1) ξ is independent of the initial energy of the scattered neutron; (2) ξ is a function only of the atomic mass of the scattering nuclide; (3) in elastic collisions, the neutron loses on average the same logarithmic fraction of its energy, regardless of its initial energy; and (4) because of this last property, we can easily calculate N, the average number of elastic scatters required to reduce a neutron of initial energy E_1 to a lower energy E_2, which is

$$N = \frac{1}{\xi} \ln\left(\frac{E_1}{E_2}\right).$$

The following table assumes $E_1 = 2$ MeV and $E_2 = 0.025$ eV. You can calculate these values using the above equations.

Material	A	α	ξ	N	Material	A	α	ξ	N
H	1	0	1	18.2	He	4	0.360	0.425	42.8
H_2O	1&16	-	0.920	19.8	Be	9	0.640	0.207	88.1
D	2	0.111	0.725	25.1	C	12	0.716	0.158	115
D_2O	2&16	-	0.509	35.7	^{238}U	238	0.983	0.0084	2172

7.16 FISSION AND ITS FUNDAMENTALS

We will start off with a **spontaneous fission**, which occurs as a form of radio-active decay via the barrier penetration mechanism of quantum mechanics. The probability of spontaneous fission occurrence is quite low. ^{235}U, ^{238}U, and ^{239}Pu undergo *spontaneous fission* to a limited extent.^{238}U, for example, yields about 60 neutrons/h/g.

Californium-252, ^{252}Cf$_{98}$, with a half-life $T^{1/2}$ of 2.645 y, has a significantly high rate of spontaneous fission. Its reaction is ^{252}Cf$_{98}$ → ^{248}Cm$_{96}$ + α with γ (yield fraction) of 96.9% of decays and spontaneous fission γ(yield fraction) of 3.1% of decays with ν of 3.76 neutrons/fission. Here, the neutron activity is roughly 2.3×10^{12} sec^{-1}g.

7.17 NEUTRON-INDUCED NUCLEAR FISSION

Figure 7.5 illustrates the general concept of neutron-induced nuclear fission. For nuclides with atomic number $Z < 90$ to fission requires excessively high excitation energies. Charged particles and γ-rays may also cause fission, but due to their low probabilities are not practical for sustaining chain reactions.

The following definitions are important and as such should be *memorized*.

- **Fissile** nuclide—Fission is possible with neutrons of *any* energy. Examples are^{235}U$_{92}$ (the only *naturally* occurring fissile nuclide), ^{233}U$_{92}$, ^{239}Pu$_{94}$, and ^{241}Pu$_{94}$.

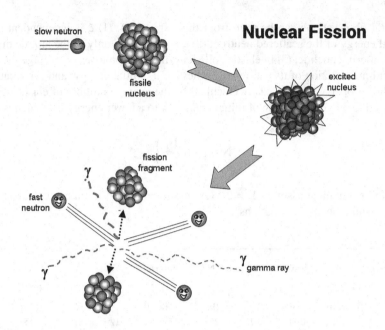

FIGURE 7.5   Sketch of nuclear fission.

- **Fissionable** nuclide—A nuclide that can undergo neutron- induced fission. These are the above *fissile* nuclides plus $^{232}Th_{90}$, $^{238}U_{92}$, $^{240}Pu_{94}$, etc.
- **Fertile** nuclide—A target nuclide for producing a *fissile* nuclide that is *not* naturally occurring—e.g., $^{232}Th_{90}$, $^{238}U_{92}$, $^{240}Pu_{94}$.

7.18   THERMAL NEUTRONS

Neutrons that are in *thermal equilibrium* with the atoms of the medium through which they are moving are termed **thermal neutrons**. Typically, thermal neutrons are those with a kinetic energy less than about 1 eV. The distribution of the speeds of thermal neutrons roughly follows a Maxwellian distribution. Thermal neutrons are as likely to gain kinetic energy as losing it from a scattering collision. Thermal neutrons have enhanced probability of causing fission. At 293 K, the most probable neutron kinetic energy is 0.025 eV, which is equivalent to a neutron speed of 2200 m/s.

7.19   FISSION PRODUCTS (FRAGMENTS)

In nuclear engineering, the definitions of fission fragments and fission products are extremely important and often being discussed in research and development. So, it is important to memorize. But what are they? Well, the immediate nuclides produced from a neutron-induced fission are termed **fission fragments**. They are unstable and the resulting daughter nuclides produced from the resulting radioactive decay chains are termed **fission products**. The fission yield plot is shown in Figure 7.6.

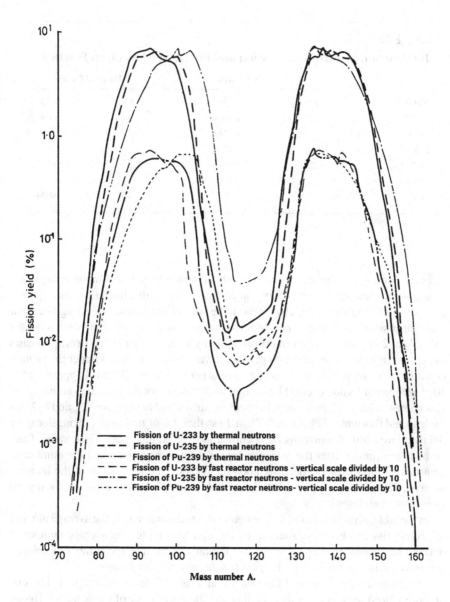

FIGURE 7.6   Yield fractions of fission fragments versus their atomic mass number.

The probability of fissions yielding fragments of equal mass increases with the energy of the incident neutron (see Figure 7.6). The valley in the curve nearly vanishes for fission caused by neutrons with energies of tens of MeV. We can see that to have a **ternary fission**, which is when three fission fragments are produced, is extremely rare, typically about 1 out of 400 cases. Also, fission fragments tend to be neutron-rich with respect to stable nuclides; therefore, they undergo β decay. Next fission fragments are both highly charged and highly energetic. They slow down

TABLE 7.1
Total Fission Neutrons per Fission and the Delayed Neutron Fraction

Nuclide	Fast Fission		Thermal Fission	
	\<v\>	β	\<v\>	β
U-233	2.62	0.0026	2.48	0.0026
U-235	2.57	0.0064	2.43	0.0065
U-238	2.79	0.0148	–	–
Pu-239	3.09	0.0020	2.87	0.0021
Pu-240	3.3	0.0026	–	–
Pu-241	–	–	3.14	0.0049

via collisions with adjacent atoms a result of which they lose kinetic energy and charge—they lose charge by picking up electrons. Generally, the half-lives of fission products earlier in the decay chain tend to be shorter than those occurring later. For a nucleus fissions, the ratio of neutrons to protons would stay the same were it not for the typical 2 or 3 neutrons given off promptly by fission—that is, the **prompt fission neutrons**. In general, the neutrons per fission vary between 0 and 8 under the prompt condition with an average number of neutrons per fission of 2.5. This happens within 10^{-14} seconds of fission event. The ratio of neutrons to protons increases above 1:1 as Z becomes larger (e.g., prominent isotopes of carbon and oxygen are $^{12}C_6$ and $^{16}O_8$ but for lead and thorium, $^{207}Pb_{82}$ and $^{232}Th_{90}$). Less than 1% of fission fragments decay by **delayed emission of neutrons** (also known as delayed neutrons). Time varies from seconds to minutes after the fission event for delayed neutrons. And the number of neutrons emitted varies depending upon the nucleus and the energy of the incident neutron. Table 7.1 provides the details of total fission neutrons per fission for several important nuclides.

In the table, \<v\> is the total fission neutrons per fission; that is, the average number of prompt fission neutrons produced by thermal fission plus the average number of delayed neutrons emitted per fission. β is the ratio of the average number of delayed neutrons emitted per fission to the total fission neutrons per fission.

The dominant decay mode of fission products is by β decay (accompanied by one or more γ-rays) which moves the daughter nuclide toward line of stable nuclei. **Decay heat** occurs when about 7% of the approximately 200 MeV produced per fission (the Q-value) is due to this β decay plus its γ-rays.

7.20 INITIAL ENERGY OF FISSION FRAGMENTS

For a fission caused by a thermal neutron (kinetic energy < 1 eV), we can ignore the incident neutron's kinetic energy. From the post-fission conservation of momentum,

$$m_L v_L = m_H v_H,$$

where L and H are the lighter and heavier fission elements, respectively. Thus, the ratio of the fission fragment kinetic energies is

$$\frac{KE_L}{KE_H} = \frac{\left(\frac{1}{2}m_L v_L^2\right)}{\left(\frac{1}{2}m_H v_H^2\right)} = \frac{m_H}{m_l}$$

Example 7.6

Consider the neutron-induced fission reaction:

$$^{235}U_{92} + {}^1n_0 \rightarrow {}^{139}Xe_{54} + ? + 2({}^1n_0) + 7(\gamma).$$

The calculated prompt energy release E_p is 183.6 MeV with the kinetic energies of the prompt neutrons and gamma rays being 5.2 MeV and 6.7 MeV, respectively. Complete the fission reaction and determine the KE of the heavier fission fragment.

Solution

We know that $235 + 1 = 139 + y + 2(1)$ and $92 + 0 = 54 + x + 2(0)$. Thus, $y = 95$ and $x = 38$. So, the isotope is strontium-95 and the reaction is

$$^{235}U_{92} + {}^1n_0 \rightarrow {}^{139}Xe_{54} + {}^{95}Sr_{38} + 2({}^1n_0) + 7(\gamma).$$

Next, we can see that $A_H = 139$ and $A_L = 95$. Thus,

$KE_L + KE_H = 183.6$ MeV – kinetic energy of the prompt neutrons (say 5.2 MeV) and gamma rays (say 6.7 MeV) = 171.7 MeV

$$\frac{KE_L}{KE_H} = \frac{\left(\frac{1}{2}m_L v_L^2\right)}{\left(\frac{1}{2}m_H v_H^2\right)} = \frac{m_H}{m_{IL}} \rightarrow KE_H = KE_L \frac{m_L}{m_H}$$

$(171.7 \text{ MeV} - KE_H)(95/139).$

So, $KE_H = \mathbf{69.8}$ **MeV**.

7.21   FISSION-NEUTRON ENERGY SPECTRUM

Figure 7.7 shows an example of uranium-235 prompt neutron fission spectrum.

Here, it is possible to describe this energy distribution using the following mathematics. First, let $\chi(E)dE$ be the average number of fission neutrons emitted with

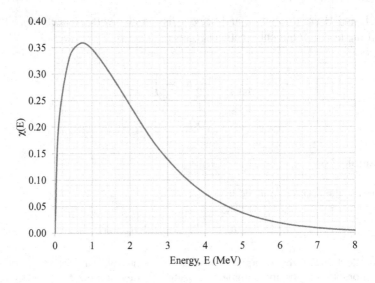

FIGURE 7.7   The energy distribution of prompt fission neutrons for U-235.

laboratory energy between E and E + dE per fission neutron. Here, $\chi(E)$ is described by the Watt distribution, which is

$$\chi(E) = ae^{-E/b} \sinh(cE)^{1/2},$$

where a, b, and c are the fitting constants.

Example 7.7

For uranium-fueled light water reactors,

$\chi(E) = 0.453 \exp(-1.036\ E) \sinh[(2.29E)^{1/2}]$ with E in units of MeV.

Estimate <E> in MeV.

Solution

Here, $\chi(E)$ reaches its maximum at ~0.7 MeV. We can use the fact that

$$\int_0^x x(E)\,dE = 1.\ \text{Thus,}\ <E> = \int_0^x Ex(E)\,dE \approx 2.0\ \text{MeV}.$$

7.22   FISSION ENERGY PRODUCTION

We already learned that the energy release from a single neutron-induced fission is roughly 200 MeV. Note that the assumption of 200 MeV produced per fission does

TABLE 7.2
Energy Release from a Nuclear Fission

Energy Source	Fission Energy (MeV)	Heat Produced MeV	% of Total	Range and Time Delay
Fission fragments	168	168	84	<mm, instantaneous
Neutrons	5	5	2.5	10-100 cm, instantaneous
Prompt γ-rays	7	7	3.5	100 cm, instantaneous
Delayed radiations				~ mm, delayed
β particles[a]	8	8	4	Delayed
γ-rays	7	7	3.5	
Neutrinos	12	0		
Radiative capture γ-rays[b]	–	3–9	2.5	100 cm, delayed
Total	207	198–204	100	

[a] Includes energy carried by β-particles and antineutrinos; the latter do not produce heat in reactors.

[b] Nonfission capture reactions contribute heat energy to all systems; design-specific considerations may change this number by a factor of two in either direction.

TABLE 7.3
Average Energy Produced in the Fission of Uranium-235 by Thermal Neutrons

	Energy from Fission (MeV)	Recoverable in Core (MeV)
Prompt:		
Kinetic energy of the fission products	168	168
Kinetic energy of prompt fission neutrons	5	5
Fission gamma rays	7	7
Gamma rays from neutron capture	–	3–9
Delayed:		
Fission product beta-decay energy	8	8
Fission product gamma-decay energy	7	7
Neutrino kinetic energy	12	0
Total energy (MeV)	207	198–204

include decay heat as shown in Table 7.2, and Table 7.3 indicates details for uranium-235 fission.

7.23 FISSION ENERGY CONVERSION METHODS

We often need to convert energy units useful at the microscopic level (e.g., MeV) to macroscopic units (e.g., kWh). Assuming that 200 MeV of *recoverable* energy is produced per fission; that is, $1 \text{ W} = (1 \text{ J/s})/[1.602 \times 10^{-13} \text{ J/MeV} \times 200 \text{ MeV/fis}]$. We can find that it takes 3.1×10^{10} fissions/s to produce approximately 1 watt of power.

Now, if A is the gram atomic weight of the fissionable isotope and N_A is Avogadro's number, then A/N_A is the mass (g) per atom of the isotope. Then, in the case of ^{235}U, the mass that must be *fissioned* to generate a power of 1 MW is

$$1 \text{ W} \rightarrow (3.1 \times 10^{10} \text{ fis/s})(86400 \text{ s/d})(235 \text{ g/mol})/(0.6022 \times 10^{24} \text{ at/mol}).$$

We derived the right-hand side of the equation from Avogadro's law, that is, $N_{atoms} = mN_A/A$ or solving for the mass m, which is $m = N_{atoms} A/N_A$.

What we want is the rate at which the mass is fissioned in units of *g/d*. In terms of rates, then we have $m' = N'_{atoms} A/N_A$. N'_{atoms} is then the number of atoms fissions per day and is equal to $N'_{atoms} = (3.1 \times 10^{10} \text{ fis/s})(86400 \text{ s/d})$.

Although *fissions* (fis) and *atoms* (at) are not dimensional type units of measurement, they essentially cancel each other in the above equation, because each fission converts one atom of ^{235}U. The units of *moles* (mol) and *seconds* (s) also cancel, so we are left with g/d of ^{235}U necessary to generate 1 W of power by fission. Solving the right-hand-side equation then yields

$$1 \text{ W} \rightarrow 1.05 \times 10^{-6} \text{ g/d OR } 1 \text{ MW} \rightarrow 1.05 \text{ g/d OR } 1 \text{ MWd} \rightarrow 1.05 \text{ g}.$$

Thus, the fission of 1 g of ^{235}U yields about 1 MWd (megawatt-day) of fission energy.

In a thermal reactor, about 85% of the neutrons absorbed result in ^{235}U fission. The remaining 15% result in radiative capture. Hence, the actual *consumption rate* of ^{235}U equals the (fission rate of ^{235}U)/0.85. Therefore, 1 MWd = 1.05 g/0.85 g of ^{235}U consumed or 1 MWd = 1.24 g of ^{235}U consumed. The first conversion factor is often used to estimate the amount of uranium-235 needed to be used for the reactor while the corrected conversion factor is being used for more precise calculation.

7.24 FUSION REACTIONS

Some of the more common fusion reactions along with the amount of kinetic energy released follow:

$$^2D_1 + {}^2D_1 \rightarrow {}^3T_1 + {}^1H_1 \qquad Q = 4.03 \text{ MeV}$$

$$^2D_1 + {}^2D_1 \rightarrow {}^3He_2 + {}^1n_0 \qquad Q = 3.27 \text{ MeV}$$

$$^2D_1 + {}^3T_1 \rightarrow {}^4He_2 + {}^1n_0 \qquad Q = 17.59 \text{ MeV}$$

$$^2D_1 + {}^3He_2 \rightarrow {}^4He_2 + {}^1H_1 \qquad Q = 18.35 \text{ MeV}$$

$$^3T_1 + {}^3T_1 \rightarrow {}^4He_2 + 2{}^1n_0 \qquad Q = 11.33 \text{ MeV}$$

$$^1H_1 + {}^6Li_3 \rightarrow {}^4He_2 + {}^3He_2 \qquad Q = 4.02 \text{ MeV}$$

$$^1H_1 + {}^{11}B_5 \rightarrow 3({}^4He_2) \qquad Q = 8.08 \text{ MeV}.$$

All the above have *threshold* energies in the range of a few keV to several hundred keV because of the repulsive Coulomb forces between the two reactants. The required kinetic energy to produce fusion can be provided by particle accelerators, but energy input to the accelerator exceeds that released by fusion.

7.25 THERMONUCLEAR FUSION

Fusion reactions are induced by the thermal motions of the reactants. The required kinetic (threshold) energies of the reactants are obtained by high temperatures. Here,

$$<E> = 3kT/2,$$

where k is the Boltzmann constant, which is 0.861735×10^{-4} eV/K^{-1} and T is the temperature that must be in the range of 10×10^6 to 300×10^6 K. The reactants exist as a *plasma*, that is, electrons and positively charged nuclei.

The strategic types for fusion reactors are: (1) gravitational confinement, (2) magnetic confinement, and (3) inertial confinement. **Gravitational confinement** is the standard solar model—the theoretical model for fusion reactions thought to power the stars. The estimated solar core conditions necessary to sustain the fusion reaction are 15×10^6 K and 4×10^{16} Pa ($\sim 400 \times 10^9$ atm).

Magnetic confinement is the type that uses magnetic fields to heat and confine plasma (Tokamak machines). It has plasma at densities of $\sim 10^{15}$ particles/cm^3. There is a difficulty in achieving stability—that is, preventing the plasma from contacting the walls of the device where it is immediately chilled to normal temperatures. Moreover, instabilities increase with increasing plasma density and temperature. The idea is to confine a plasma with $<E> = 10$ keV/particle and a density equal to that of the atmosphere ($\sim 10^{19}$ particles/cm^3), which requires magnetic pressures of greater than 10^5 atm. Reactor chamber walls must be cooled since they are heated by fusion products. In addition, fusion neutrons must be shielded and fusion heat must be converted to electricity. Here, the most promising fusion reaction, that is, the one requiring the lowest plasma temperature to overcome the Coulomb barrier, is

$$^2D_1 + {}^3T_1 \rightarrow {}^4He_2 \ (3.54 \text{ MeV}) + {}^1n_0 \ (14.05 \text{ MeV}) \text{ with } Q = 17.59 \text{ MeV}.$$

Inertial Confinement used a small pellet of the reactants to rapidly heat a plasma and compress to high density via simultaneous bombardment from several directions with powerful laser pulses. It uses the same principle as for thermonuclear bombs where heat and compression are provided by a small fission (atomic) bomb. A simpler type of inertial confinement is inertial electrostatic confinement (IEC) where a gas such as deuterium is introduced into an evacuated pressure chamber. A high voltage is placed across the chamber so that the outer shell acts as an anode and a metallic grid placed at the center acts as a cathode. The deuterium atoms are ionized and the positively charged nuclei are accelerated toward the cathode where some collide with enough energy to overcome the Coulombic repulsion barrier and fuse. The VCU fusor is an example of an IEC device (see Figure 7.8).

FIGURE 7.8 (Left) "Star mode" inside the VCU IEC fusor. (Right) VCU IEC fusor.

Important Terms (in the Order of Appearance Showing the Frame Number)

Conservation quantity	Thermal neutron	Fission fragments
Peripheral process	Neutron moderation	Fission products
Compound nucleus	Neutron thermalization	Ternary fission
Virtual state	Isotropic scatter	Prompt fission neutrons
Bound state	Lethargy	Delayed emission neutrons
Exit channel	Slowing-down decrement	Decay heat
Exoergic reaction	Spontaneous fission	Gravitational confinement
Endoergic reaction	Fissile nuclide	Magnetic confinement
Kinematic threshold energy	Fertile nuclide	Inertial confinement
Threshold energy	Fissionable nuclide	

BIBLIOGRAPHY

Bush, H.D., *Atomic and Nuclear Physics,* Prentice Hall, Englewood Cliffs, NJ, 1962.

Coombe, R.A., *An Introduction to Radioactivity for Engineers.* Macmillan Publishing Co., Inc., New York, 1968.

Coppi, B., and Rem, J., "The Tokamak Approach in Fusion Research," *Scientific American,* **229** (2), 1974, pp. 50–59.

Glasstone, S., and R.H. Loveberg, *Controlled Thermonuclear Reactions.* Princeton, N.J.: D. Van Nostrand Company, 1960.

Keepin, G.R., *Physics of Nuclear Kinetics.* New York: Addison-Wesley, 1965.

Lamarsh, J.R., and Baratta, A.J., *Introduction to Nuclear Engineering,* 3rd Ed., Prentice Hall, Englewood Cliffs, NJ, 2001.

Mayo, R.M., *Nuclear Concepts for Engineers*, American Nuclear Society, La Grange Park, IL, 1998.

Shultis, J.K. and R.E. Faw, *Fundamentals of Nuclear Science and Engineering 2nd Ed.,* CRC Press, Taylor & Francis Group, Boca Raton, FL, 2008.

FURTHER EXERCISES

A. True or False: If the statement is false, give a counterexample or explain the correction. If the statement is true, explain why it is true.
1. Fission fragments are not radioactive.
2. There is a high U-235 fission yield for fission fragments with mass number of 110–120.
3. The energy released per fission of U-235 is approximately 200 MeV.
4. In the energy distribution for fission induced by thermal neutrons in U-235, neutrino energy provides a significant part as the heat source.
5. Exoergic reaction occurs when Q-value is greater than zero.
6. Two types of reaction threshold energies that can be encountered are Kinematic and Thermodynamic thresholds.
7. Lethargy of helium-4 is larger than that of uranium-238.
8. Radiative capture is highly hazardous in nature.
9. Prompt neutrons are emitted within 10^{-5} seconds of fission event.
10. Delayed neutrons are typically about 5% for the thermal fission, which were generated from the neutron decay of fission products.

B. Problems
1. Lithium-7 is quite popular for having several exit channels after being induced by thermal neutrons. For each of the following possible reactions, all of which create the compound nucleus ^{7}Li,

$$^1n_0 + {}^6Li_3 \rightarrow ({}^7Li_3)^* \rightarrow {}^7Li_3 + \gamma$$

$$^1n_0 + {}^6Li_3 \rightarrow ({}^7Li_3)^* \rightarrow {}^6Li_3 + {}^1n_0$$

$$^1n_0 + {}^6Li_3 \rightarrow ({}^7Li_3)^* \rightarrow {}^6He_2 + p$$

$$^1n_0 + {}^6Li_3 \rightarrow ({}^7Li_3)^* \rightarrow {}^5He_2 + d$$

$$^1n_0 + {}^6Li_3 \rightarrow ({}^7Li_3)^* \rightarrow {}^3H_1 + \alpha$$

calculate (a) the Q-value, (b) the kinematic threshold energy, and (c) the minimum kinetic energy of the products. Summarize your calculations in a table. Note that the minimum kinetic energy imparted to the

products is the sum of the Q-value and the applicable kinematic threshold energy should the reaction be endothermic.

2. An important radionuclide produced in water-cooled nuclear reactors is
    ^{16}N which has a half-life of 7.13 s and emits very energetic gamma rays
    of 6.1 and 7.1 MeV. This nuclide is produced by the endoergic reaction
    $^{16}O(n, p)^{16}N$. What is the minimum energy of the neutron needed to pro-
    duce ^{16}N? Note: Beware of red herrings, that is, data that are given but
    not needed in solving the problem.

3. In nuclear medical application for diagnoses of tumors, it is possible to
    produce the isotope ^{18}F by irradiating lithium carbonate (Li_2CO_3) with
    neutrons. First, the neutrons interact with ^6Li to produce tritons (nuclei
    of ^3H), which in turn, interact with the oxygen (^{16}O) to produce ^{18}F. (a)
    What are the two nuclear reactions? (b) Calculate the Q-value for each
    reaction in MeV. (c) Calculate the threshold energy for each reaction
    in MeV. (d) Can thermal neutrons (i.e., neutrons of very low kinetic
    energy, less than 1 eV) be used to create ^{18}F? If so, show why.

4. How many elastic scatters, on the average, are required to slow down a
    1-MeV neutron to below 1 eV in (a) ^{16}O and in (b) ^{56}Fe?

5. How many neutrons per second are emitted spontaneously from a 1 mg
    sample of ^{252}Cf? For ^{252}Cf, assume the following data:

    $T_{1/2} = 2.745$ y, $\gamma_n = 0.0309$ (i.e., fraction of decays that are sponta-
    neous fission), and

    $\nu = 3.73$ (average number of neutrons per fission)

6. An experimentalist observes four prompt neutrons are being produced
    with one fission fragment being ^{121}Ag from a neutron-induced fission
    reaction of uranium-235. (a) What is the other fission fragment? (b) How
    much energy is liberated promptly (i.e., before the fission fragments
    begin to decay; in other words, what is the Q-value)?

7. Calculate the Coulombic threshold energies in MeV for the following
    fusion reactions.

    $$^2D_1 + {}^2D_1 \rightarrow {}^3T_1 + {}^1H_1$$

    $$^2D_1 + {}^3T_1 \rightarrow {}^4He_2 + {}^1n_0$$

    $$^3T_1 + {}^3T_1 \rightarrow {}^4He_2 + 2({}^1n_0)$$

    $$^1H_1 + {}^{11}B_5 \rightarrow 3({}^4He_2)$$

8. The Q-value for the reaction $^{14}C(p, n)^{14}N$ is -0.6259 MeV (i.e., a **negative**
    0.6259 MeV). Here, p represents a proton and n a neutron. What is the
    threshold energy in MeV for the reaction?

9. Given the nuclear fission reaction: $^{235}U_{92} + {}^1n_0 \rightarrow {}^{139}Xe_{54} + {}^{95}Sr_{38} + 2({}^1n_0) + 7(\gamma)$,
   a. Calculate the Q-value (MeV) for the nuclear fission reaction.
   b. Assuming that the calculated Q-value from part (a) is the average amount of useful energy released per fission, how many fissions per second are required to produce 200 MW (2×10^8 W) of power?
   c. If 85% of the neutrons absorbed in ^{235}U fission, with the remaining 15% resulting in radiative capture, how many grams per second of U-235 are consumed in producing 200 MW of power?

8 Radiation Interactions with Matter

OBJECTIVES

After studying this chapter, the reader should be able to:

1. Gain knowledge on the concept of a cross section that quantifies the probability of the interaction of radiation with the atoms of a material, especially photon and neutron interactions.
2. Understand the significance of radiation flux and radiation current.
3. Derive the transport of radiation and its associated weakening due to geometric attenuation and material attenuation.
4. Calculate and estimate the radiation interaction rates.
5. Understand the fundamentals of Bremsstrahlung, charged particles, straggling, specific ionization, stopping power, and linear energy transfer.

8.1 NUCLEAR RADIATION ENVIRONMENT

There are about 10^9 photons per nucleon in the universe at large. Here, neutrino flux is about $10^9/cm^2$-s. To produce biological damage, radiation must ionize cellular atoms, which in turn alters molecular bonds and changes cell chemistry. To produce damage in structural and electrical materials, radiation must cause interactions that disrupt crystalline and molecular bonds. One possible classification of radiation is as directly ionizing or indirectly ionizing. **Directly ionizing radiation** exists when charged particles interact with the Coulomb field of atoms. Examples are α particles, β particles, and fission fragments. **Indirectly ionizing radiation** happens when neutral particles that through their interactions produce charged particles known as **secondary radiation.** Examples are neutrons and photons (i.e., X-rays and gamma-rays). These interactions are dominated by short-range forces.

8.2 MICROSCOPIC CROSS SECTIONS

Cross sections relate to the *probability* of interactions between radiation (e.g., neutrons, electrons, photons) and the material they pass through. We will begin the discussion of microscopic cross sections with neutrons since neutrons are responsible for neutron-induced nuclear fission and, being uncharged, are not influenced by electromagnetic forces.

Consider a single neutron *normally* incident on a *thin* target disk—say one atomic layer thick so that *no atom in the target is shielded by another atom* (see Figure 8.1). Let N be the number of target atoms per cm³ i.e., the *atom density* or

DOI: 10.1201/9781003272588-8

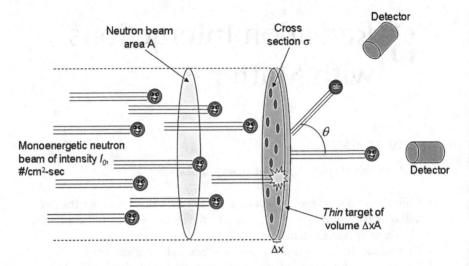

FIGURE 8.1   Schematic of a method for measuring a microscopic cross section.

atom number density), dA be the area of target (cm^2), dx be the thickness of target (cm), and σ be the *cross-sectional target* area presented by a *single* nucleus. Then, the total area of target is

Total area of target = (# of nuclei in disk) × (area per nucleus) = N dA dx σ.

The probability that a neutron will interact in traveling distance dx through the disk is

(total target area)/(disk surface area) = (N dA dx σ)/dA = Nσdx.

Let σ equal the interaction probability divided by (Ndx); that is, the interaction probability per unit atom density per unit distance of neutron travel. Then, the unit of σ is the unit of area, for example, cm^2. Thus, σ is **the microscopic cross section**. Basically, the microscopic cross section characterizes the probability of a neutron–nuclear reaction for the nucleus (see Figure 8.2).

σ is a function of (1) the neutron speed or energy (there is some dependence on the angle of collision, but typically it is so weak that it can be ignored), (2) the target nucleus, and (3) the type of interaction (e.g., absorption, scattering, inelastic scattering).

If the neutron–nucleus interaction could be visualized as a classical collision, σ would be the effective *cross-sectional area* presented by the nucleus to the neutron beam. Since the diameter of a nucleus is about 10^{-12}cm, the geometric cross-sectional area presented by a typical nucleus would be approximately the square of its diameter; that is, $\approx 10^{-24}$cm^2. We, therefore, define the unit of measurement for the microscopic cross section as the barn, abbreviated as b, where 1 barn = 1 b = 10^{-24}cm^2.

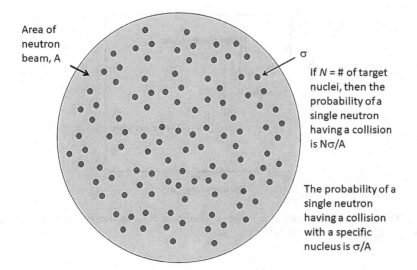

FIGURE 8.2 Definition of the microscopic cross section, a "thin" target from approaching neutron's point of view. Note: N represents the number of atoms, whereas in the derivations and equations in the text, N represents the atom number density.

8.3 NEUTRON ENERGY DEPENDENCE OF THE MICROSCOPIC CROSS SECTIONS

As depicted in the following simplified plot for a heavy nucleus (see Figure 8.3), plots of the neutron microscopic cross sections as a function of neutron energy can be subdivided into three important regions. The first region is known as the **thermal energy region** (also known as the **1/v absorption region**). Here, the microscopic cross section varies as the inverse of the neutron speed (i.e., as 1/v). This is due to the fact that slower neutrons spend more time in the vicinity of a nucleus and thereby experience nuclear forces for a longer time. This region is also termed the *thermal energy* region in that the neutrons are approximately in thermal equilibrium with the material. The high neutron energy cutoff for the 1/v absorption region is about 1 eV. The second region is the **resonance absorption region**. To the right of the 1/v absorption region, that is, starting at about 1 eV neutron energy, peaks in the value of the microscopic absorption cross section, called **resonances**, occur when the neutron energies match the energy differences between the nucleus' discrete quantum levels as described by the nucleus shell model, that is, they correspond to the formation of an excited state of the resulting compound nucleus. The third region is the **fast energy** region. Above ~10 keV neutron energy, the absorption cross sections show relatively little variation with neutron energy.

In analogy to the neutron absorption regions, there are neutron scattering energy regions described as follows. First is the **potential scattering** region. In this region, the microscopic cross sections show little change with increasing neutron energy. Instead of penetrating the nucleus, the neutron scatters elastically off the nuclear

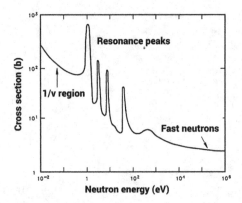

FIGURE 8.3 Simplified plot of the neutron absorption cross section versus neutron energy for a fissile nucleus such as U-235 or Pu-239.

potential similar to the classical "billiard ball" collision described earlier in the course. The microscopic cross sections show little change in the energy range of 1 eV up to MeV. The neutron scattering cross section $\sigma_s = 4\pi R^2$, where R is the radius of the nucleus, equal to approximately $1.25 \times 10^{-13} A^{1/3}$ cm, and A is the atomic mass number of the nucleus. Second is the **resonance scattering** region. The cross section behavior in this region is similar to that in the resonance absorption region.

Example 8.1

Extrapolation of the value of a microscopic cross section in the thermal region. The microscopic radiative capture cross section of U-233 at a neutron energy of 0.0253 eV is 46.9 barns. What is the value of the cross section at a neutron energy of 0.01 eV?

Solution

In the thermal energy region, neutron absorption cross sections vary as 1/v, where v is the neutron speed. Since the neutron kinetic energy is given by the classical expression, $E = \frac{1}{2} m_n v^2$, we note that the square root of the neutron energy is proportional to the speed. Therefore,

$$\frac{\sigma_1}{\sigma_2} = \frac{v_2}{v_1} \text{ and since } v \propto \sqrt{E} \rightarrow \frac{\sigma_1}{\sigma_2} = \left[\frac{E_2}{E_1}\right]^{\frac{1}{2}} \rightarrow \sigma_1 = \sigma_2 \left[\frac{E_2}{E_1}\right]^{\frac{1}{2}}$$

$$\sigma_1 = (46.0 b) \left[\frac{0.0253\,eV}{0.01\,eV}\right]^{\frac{1}{2}} = (46.0 b)(1.59) = \mathbf{72.2\,b}$$

The following plots (Figures 8.4–8.8) of microscopic cross sections from various cross section libraries are available at: http://atom.kaeri.re.kr/endfplot.shtml.

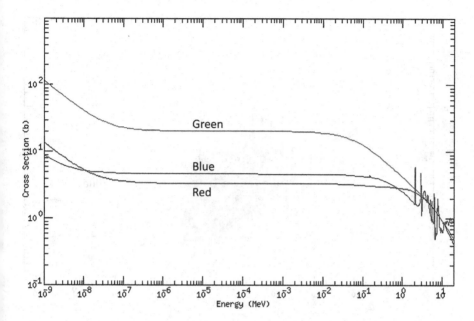

FIGURE 8.4   Hydrogen (red), deuterium (green) and oxygen (blue) scattering cross sections.

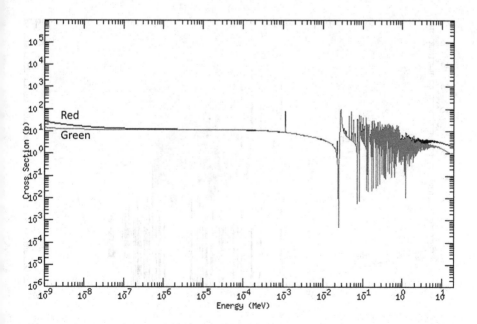

FIGURE 8.5 ^{56}Fe cross sections: total (red) and elastic (green).

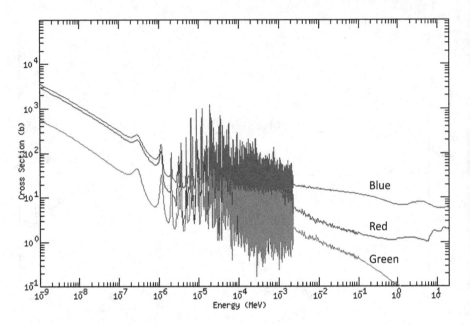

FIGURE 8.6 ^{235}U cross sections: total (blue), fission (red), radiative capture (green).

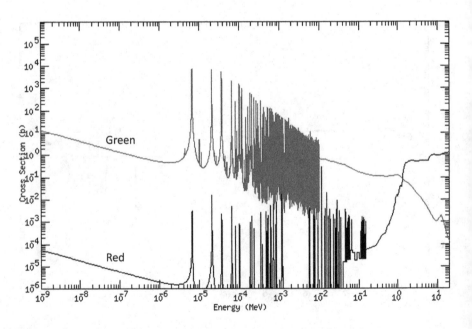

FIGURE 8.7 ^{238}U fission (red) and radiative capture (green) cross sections.

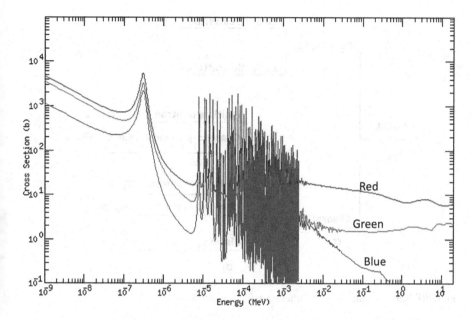

FIGURE 8.8 ^{239}Pu cross sections: total (red), fission (green) and radiative capture (blue).

Hydrogen displays no resonance peaks since the hydrogen nucleus, a proton, possesses no internal structure (see Figure 8.4). It should be noted that the appearance of resonance scattering regions starts to occur for deuterium and oxygen at higher neutron energies (see Figure 8.4). Since heavier nuclei display resonance structure, *as the atomic weight of the nuclei increases, the resonance structure becomes more complex and moves to the left on the neutron energy scale.* This behavior is apparent in the following figures (Figures 8.5–8.8) for both scattering and absorption cross sections.

Again, the resonance regions are moving to the left with higher atomic weight. Recall from the previous chapter notes that *radiative capture* is the reaction where a neutron is absorbed and a gamma-ray is released. Again, cross section data are obtained from ENDF-VI using ENDFPLOT from the Korean Atomic Energy Research Institute.

8.4   SCHEMATIC OF MICROSCOPIC CROSS SECTION

At *a specific incident neutron speed* (i.e., energy), the various microscopic cross sections are additive. Here, the microscopic total cross section (σ_t) is the sum of the microscopic scattering cross section (σ_s) and microscopic absorption cross section (σ_a). Moving further down in the neutron interaction hierarchy, we have σ_e = microscopic elastic scattering cross section, σ_i = microscopic inelastic scattering cross section, σ_f = microscopic fission cross section, and σ_γ = microscopic radiative capture cross section. From that nature, $\sigma_s = \sigma_e + \sigma_i$ and $\sigma_a = \sigma_f + \sigma_\gamma + \sigma_\alpha + \sigma_p + \sigma_{2n}$ + so on. This can be easily explained by Figure 8.9.

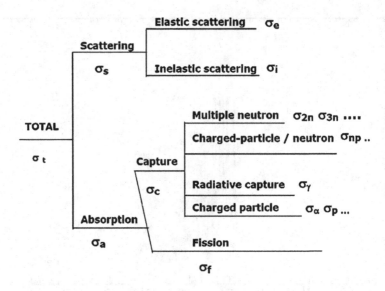

FIGURE 8.9   Schematic of the microscopic cross sections.

Example 8.2

What is the probability that a thermal neutron that *interacts* with a ^{235}U nucleus is *absorbed*?

Solution

Well, the answer is $p_{abs} = \sigma_a/\sigma_t = (\sigma_f + \sigma_\gamma)/\sigma_t$.

Example 8.3

What is the probability that the absorption of a thermal neutron by a ^{235}U nucleus results in *fission*?

Solution

Here, the answer is easily expressed as $p_{abs} = \sigma_f/\sigma_a = \sigma_f/(\sigma_f + \sigma_\gamma)$.

8.5   ATTENUATION OF RADIATION BEAM

Microscopic cross sections are necessary for the calculation of **macroscopic cross sections,** denoted by Σ. For a given nuclide, the associated macroscopic cross section is the product of its microscopic cross section and the atom number density of the nuclide, yields the unit of cm^{-1}.

$$\Sigma_t = N\sigma_t$$

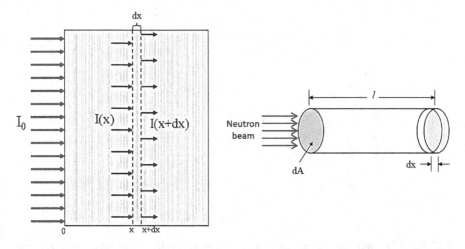

FIGURE 8.10   (Left) Collimated neutron beam impinging on material. (Right) Schematic of the interaction of a neutron beam through a section of material used in the derivation of the macroscopic cross section.

It is extremely important to note that the **macroscopic cross section** Σ_i characterizes the probability of radiation interactions of type i in a chunk of material, whereas the **microscopic cross section** σ_i characterizes the probability of interaction with only a single nucleus or atom. Another important note is that although we specify neutrons as the type of radiation in the following discussion, the general concept applies to the interaction of all types of radiation with matter.

Now, we can introduce a concept behind radiation **attenuation**, which is the act of thinning or weakening, i.e., weakening in force or intensity. Consider the attenuation of a *monoenergetic* neutron beam normally incident on a *thick* target as shown in Figure 8.10. Here, I(x) is *virgin* beam intensity at any point x in the target, where *virgin* implies a neutron that has not yet interacted in any way; that is, either by absorption or scattering, with a target nucleus. This is also termed the *uncollided* beam. Unit of beam intensity is $cm^{-2}\text{-}s^{-1}$, that is, number of radiation particles per unit area per unit time. We will combine this idea along with the derivation of the macroscopic cross section to explain the attenuation.

Let I(x) = the number of particles perpendicular to and passing through cross-sectional area dA per unit time, and $dN_A = Ndx$ = number of target nuclei per cm^2 in dx, where N is the atom number density of the material. Thus, the total reaction rate per unit area in dx is therefore equal to

$$\sigma_t I \times dN = \sigma_t I N dx.$$

Equating the reaction rate with the *decrease* in beam intensity between x and x + dx,

$$-dI(x) = -[I(x + dx) - I(x)] = \sigma_t I(x) \times Ndx,$$

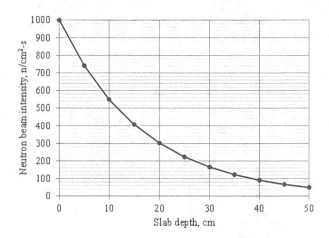

FIGURE 8.11 Material attenuation of a particle beam incident on an infinite slab showing the *uncollided* beam intensity (solid line) and the *actual* beam intensity (dashed line) versus depth of particle penetration.

which translates into the following differential equation:

$$\frac{dI}{dx} = -N\sigma_t \times I(x).$$

Does the form of this differential equation and its solution look familiar? It is equivalent to that for simple radioactive decay, that is, $dN(t)/dt = -\lambda t$ with time t replacing distance x. If the number of atoms at time zero is N_0, then the solution is $N = N_0 \exp(-\lambda t)$. Now, let $I(x=0) = I_0$, then $I(x) = I_0 \exp(-N\sigma_t\, x)$ or $I(x) = I_0 \exp(-\Sigma_t\, x)$ by utilizing the fact that $\Sigma_t = N\sigma_t$.

In analogy to the radiation decay equation, the macroscopic cross section Σ_t replaces the decay constant λ, and the particle beam intensity, I, replaces the number of atoms or atom number density N and the I can be plotted as shown in Figure 8.11.

Here, Σ_t provides the fractional change in the *uncollided neutron* beam intensity occurring over a distance *dx*, which is $\Sigma_t = (-dI/I)/dx$ (see Figure 8.12). So, we can view Σ_t *as the probability per unit path length traveled that a particle will undergo an interaction.* The ratio of the beam intensity at a depth x into the slab I(x) to the incident beam intensity I_0, $I/I_0 = \exp(-\Sigma_t x)$ gives the *probability that a particle moves a distance x without any interaction.*

8.6 MEAN FREE PATH AND ITS DERIVATION

Let p(x)dx be the probability that a particle will have its *first interaction* in dx. This is equal to the probability that the particle survives up to x without interaction *times* the probability that it does interact in the additional distance dx. Since Σ_t is the probability of interaction per path length,

$$p(x)dx = \exp(-\Sigma t x) \times \Sigma t\ dx.$$

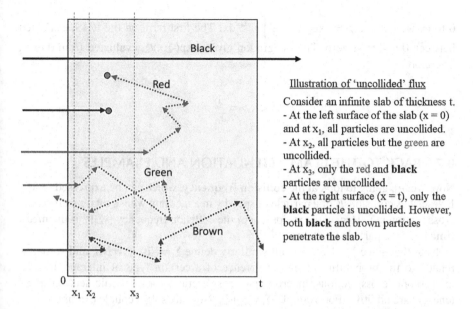

FIGURE 8.12   Infinite slab illustration of the concept of "uncollided" flux.

Carefully examined, it is possible to show that $\Sigma_t \exp(-\Sigma_t x)dx$ represents the *probability that a neutron has its first interaction in* dx. Thus, the average distance a particle travels before interacting with a nucleus $<x>$ will be

$$< x >= \int_0^\infty x\, p(x)dx = \Sigma_t \int_0^\infty x\, \exp\left(-\Sigma_t\, tx\right)dx = \frac{1}{\Sigma_t}.$$

This expression is known as the **mean free path**,

$$<x> = \lambda = \text{mean free path} = 1/\Sigma_t.$$

Here, the mean free path is usually represented by λ, we should **not** to be confused with the decay constant or photon wavelength. But, how do we jump from one expression to the final expression? Here, we will show the details of the derivation that the reciprocal of the macroscopic cross section is the mean free path. We first will start with

$$< x >= \int_0^\infty x\, p(x)dx = \Sigma_t \int_0^\infty x\, \exp\left(-\Sigma_t\, tx\right)dx.$$

Integrate the far right equation by parts; that is, we will use: $\int u\,dv = uv - \int v\,du$, where $u = x$ and $dv = \Sigma_t \exp(-\Sigma_t x)dx$. Hence, $du = dx$ and $v = -\exp(-\Sigma_t x)$. Integrating from

0 to ∞, we have $<x> = -xe^{-\Sigma_t x} |_0^\infty + \int e^{-\Sigma_t x} dx$. The first term on the RHS, evaluated between 0 and ∞, is zero. The integration gives $-\exp(-\Sigma_t x)/\Sigma_t$ evaluated from 0 to ∞. Therefore,

$$< x > = \frac{1}{\Sigma_t}.$$

8.7 BACK TO THE BEAM ATTENUATION AND EXAMPLES

Now, we are going to look at the **collision frequency**, which can be expressed by $v\Sigma_t$. It describes collisions/s of a particle of speed v in a medium whose total macroscopic cross section is Σ_t. Hence, the reciprocal of the collision frequency $[v\Sigma_t]^{-1}$ is the **mean time** between particle reactions.

Note that since $\Sigma_t = N\sigma_t$, we can similarly define $\Sigma_f = N\sigma_f$, $\Sigma_s = N\sigma_s$ and so on, as related to the probability of the occurrence of a certain type of interaction. Like microscopic cross sections, macroscopic cross sections for a specific neutron speed (energy) are additive. For example, $\Sigma_t = \Sigma_s + \Sigma_a$. Now, let's do a couple examples.

Example 8.4

A beam of monoenergetic neutrons is perpendicularly incident on an infinite slab of thickness 5.0 cm. The total macroscopic cross section for the slab is 1.2 cm^{-1}. What fraction of the neutrons in the beam that penetrate the slab without undergoing an interaction?

Solution

The fraction of the neutrons that penetrate the slab without undergoing any sort of interaction is given by

$$\frac{I_{unc}}{I_0} = \exp(-\Sigma_t X).$$

where x is the slab thickness. Thus,

$$\frac{I_{unc}}{I_0} = \exp[-(1.2 \text{ cm}^{-1})(5 \text{ cm})] = \exp(-6.0) = \textbf{2.48} \times \textbf{10}^{-3}.$$

Example 8.5

A beam of monoenergetic neutrons of intensity 9.5×10^{22} cm^{-2}-s^{-1} is perpendicularly incident on an infinite slab composed of two homogeneous materials. The material in the first part of the slab has a thickness of 2.0 cm and a total macroscopic cross section of 0.66 cm^{-1}. This is bounded by a second material of thickness 2.90 cm with a total macroscopic cross section of 0.25 cm^{-1}. What is the *uncollided* beam intensity of the neutrons that penetrate the slab? How would the answer changed if the order of the two materials in the slab was switched?

Solution

$$I_{unc} = I_0 \exp(-\Sigma_{t1}x_1)\exp(-\Sigma_{t2}x_2) = I_0\exp(-\Sigma_{t1}x_1 - \Sigma_{t2}x_2)$$

$$I_{unc} = (9.5\times10^{22}\,\text{cm}^{-2}\text{s}^{-1})\exp\left[-(0.66\,\text{cm}^{-1})(2.0\,\text{cm}) - (0.25\,\text{cm}^{-1})(2.90\,\text{cm})\right]$$

$$I_{unc} = (9.5\times10^{22}\,\text{cm}^{-2}\text{s}^{-1})\exp[-1.32 - 0.725]$$

$$= (9.5\times10^{22}\,\text{cm}^{-2}\text{s}^{-1})(0.129) = \mathbf{1.23\times10^{22}\,cm^{-2}s^{-1}}$$

With regard to the *uncollided* flux, switching the order of the slabs would have **no** impact. In reality, the actual flux would vary if the order was switched.

8.8 CROSS SECTION OF MIXTURES AND MOLECULES

For mixtures of materials, the macroscopic cross section is defined as

$$\Sigma = \sum_{i=1}^{n}\Sigma_i = \sum_{i=1}^{n}N_i\sigma_i,$$

where N_i and Σ_i are the number density and macroscopic cross section, respectively, of species i in the material. For example, consider a homogeneous mixture of two nuclear species X and Y, containing N_X and N_Y atoms/cm³ of each type. The respective microscopic cross sections are σ_X and σ_Y. The probability per unit path that a neutron interacts with the nucleus of the first type is $\Sigma_X = N_X\sigma_X$, and with the second is $\Sigma_Y = N_Y\sigma_Y$. Hence, the total probability per unit path length that a neutron interacts with *either* nucleus is $\Sigma = \Sigma_X + \Sigma_Y = N_X\sigma_X + N_Y\sigma_Y$.

If the nuclei compose atoms of a molecule, the above equation can be used to define an equivalent cross section for the molecule. This is done by simply dividing the macroscopic cross section of the mixture by the number of molecules per unit volume. If, for instance, the molecular formula is X_mY_n, the resulting cross section for the molecule is

$$\sigma = m\sigma_x + n\sigma_y.$$

Note that this set of equations $\Sigma = \sum_{i=1}^{n}\Sigma_i = \sum_{i=1}^{n}N_i\sigma_i$ is based on the assumption that the nuclei i act independently of one another when they interact with neutrons. *In some cases, particularly for low-energy elastic scattering by molecules and solids, this assumption is not valid and the above relationship does not apply.* This is illustrated in the graph shown in Figure 8.13 for water.

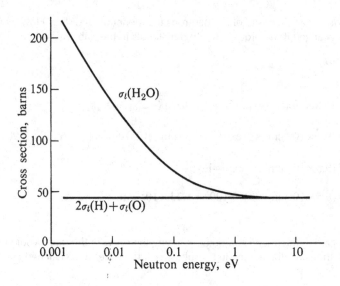

FIGURE 8.13 Water is an example where the thermal neutron microscopic cross section is <u>not</u> the atom-weighted sum of the cross sections of its constituent atoms.

Monoenergetic neutron collisions in a heterogeneous medium composed of three different isotopes X, Y and Z with number densities N_X, N_Y and N_Z, respectively:

$$\Sigma_t = N_X\sigma_t^X + N_Y\sigma_t^Y + N_Z\sigma_t^Z.$$

Example 8.6

Find the mean free path for a thermal neutron in graphite.

Solution

For carbon, $\sigma_s = 4.8$ b, $\sigma_a = 4.0 \times 10^{-3}$ b, $\rho = 1.60$ g/cm³. Therefore, the atomic number density of carbon is $N = N_a\rho/A$, where N_a is Avogadro's number and A is the gram atomic weight of carbon, which is 12 g/mol.

$$N = (0.6022 \times 10^{24} \text{ at/mol})(1.60 \text{ g/cm}^3)/(12 \text{ g/mol}) = 8.03 \times 10^{22} \text{ at/cm}^{-3}$$

$$\Sigma_t = N\sigma_t = (8.03 \times 10^{22} \text{ cm}^{-3})(4.8 \text{ b} + 4.0 \times 10^{-3} \text{ b})(10^{-24} \text{ cm}^2/\text{b}) = 0.385 \text{ cm}^{-1}$$

$$\lambda = \text{mfp} = 1/\Sigma_t = 1/0.385 \text{ cm}^{-1} = \mathbf{2.6 \text{ cm}.}$$

The ratio of the scattering to absorption cross sections that a neutron in graphite will make an average of $\sigma_s/\sigma_a = 4.8/0.004 = 1200$ scattering collisions before being absorbed.

8.9 MACROSCOPIC CROSS SECTION CALCULATION

Problems in reactor theory and radiation shielding often involve the calculation of macroscopic cross sections Σ for a homogeneous material composed of a mixture of nuclides or elements. Hopefully, this document will remove some of the confusion surrounding the procedure for finding Σ. The confusion usually results from how the mixture is characterized. Typically, the composition is described in one of three different ways: (1) w—weight (or mass) fractions (or percents), (2) γ—atom fractions (or percents), and (3) vf—volume fractions (or percents). Here, we are specifically addressing macroscopic cross sections for neutrons. Instead of referring to the different nuclides or elements that make up the homogeneous material of interest, we will refer to the different species that make up the homogeneous material, where species can refer to either a nuclide or an element depending on the specific case. There are several variables to be considered; these are: ρ_i is the material mass density for a material composed solely of species i (g/cm^3), n is the number of different species in the material, A_i is the gram atomic weight of species i (g/mol), σ_i is the microscopic cross section of species i (barns or cm^2), w_i is the weight (mass) fraction of species i in the material, γ_i is the atom fraction of species i in the material, vf_i is the volume fraction of species i in the material, Σ_i is the macroscopic cross section for a material composed solely of species i (cm^{-1}), and Σ is the macroscopic cross section for the homogeneous material (cm^{-1}).

Given the composition in **weight (mass) fractions**: $\Sigma = \rho N_a \sum_{i=1}^{n} w_i \frac{\sigma_i}{A_i}$

Given the composition in **atom fractions**: $\Sigma = \rho \frac{N_a}{A} \sum_{i=1}^{n} \gamma_i \sigma_i$ where $A = \sum_{i=1}^{n} \gamma_i A_i$

Given the composition in **volume fractions**: $\Sigma = \sum_{i=1}^{n} vf_i \Sigma_i$ where $\Sigma_i = \rho_i \frac{N_a}{A_i} \sigma_i$.

It should be noted that unlike in the equation for the mixture given previously where Σ_i was the macroscopic cross section of species *i* in the material; here, Σ_i is the macroscopic cross section of species i of a pure material.

Example 8.7

What is the macroscopic absorption cross section for thermal neutrons in water? The relevant absorption cross sections are: Hydrogen-1: $\sigma_a = \sigma_\gamma = 333$ mb $= 0.333$ b and Oxygen-16: $\sigma_a = \sigma_\gamma = 190$ µb $= 1.90 \times 10^{-5}$ b.

Solution

For water, H_2O, we know the relative abundances of the hydrogen and oxygen atoms. Therefore, $\gamma_H = 2$ and $\gamma_O = 1$. That is, two of the atoms in a water molecule are (light) hydrogen, $^1H_1$, and one is oxygen, $^{16}O_8$. Here, we've ignored the remaining hydrogen and oxygen isotopes because of their insignificant abundance.

Using the equations from above for "Calculating Macroscopic Cross Sections" given the atoms fractions from above, $\Sigma = \rho \dfrac{N_a}{A} \sum\limits_{i=1}^{n} \gamma_i \sigma_i$, where $A = \sum\limits_{i=1}^{n} \gamma_i A_i$, we first calculate the equivalent gram molecular weight, A, for the molecule.

$$A_H = (2)(1 \text{ g/mol}) = 2 \text{ g/mol} \qquad\qquad A_O = (1)(16 \text{ g/mol}) = 16 \text{ g/mol}$$

Therefore, the equivalent molecular weight is, $A = A_H + A_O = 2 + 16 \text{ g/mol} = 18 \text{ g/mol}$. Assuming a density of water of 1 g/cm³, we have for the total macroscopic absorption cross section,

$$\Sigma_a = [(1 \text{ g/cm}^3)(0.6022 \times 10^{24} \text{ at/mol})/(18 \text{ g/mol})] \ [(2/3)(0.333 \text{ b}) + (1/3)(1.90 \times 10^{-5} \text{ b})]$$

$$\Sigma_a = (3.346 \times 10^{22} \text{ cm}^{-3})(0.6333 \text{ b})(10^{-24} \text{ cm}^2/\text{b}) = \mathbf{0.0212 \ cm^{-1}}.$$

Example 8.8

Calculate the total thermal macroscopic cross section for 8 **a/o** enriched UO_2 that is mixed in a 1:3 volume ratio with graphite by using the following given data sets.

Data: density of $UO_2 = 11.0$ g/cm³, density of carbon $= 1.6$ g/cm³, total microscopic cross section of uranium-235 $= 700$ b, total microscopic cross section of uranium-238 $= 12.2$ b, thermal microscopic cross section of oxygen $= 4.03$ b, and thermal microscopic cross section of carbon $= 4.75$ b.

Solution

Because oxygen and carbon are predominantly ^{16}O and ^{12}C, respectively, we'll use the thermal microscopic cross sections for those nuclides. The two mass densities are for the pure materials and not for the mixture. Note that the 8 a/o value for the enriched UO_2 refers only to the fraction of uranium atoms that are U-235 and not to the entire UO_2 fuel pellet. The remaining atoms in the uranium (92 a/o) are assumed to be U-238. (*Note:* Usually enrichments are in w/o, weight-percent, and the relative atomic weights of the isotopes would be taken into account.)

Since all constituents of the UO_2 are provided based on atom fractions, we'll calculate the macroscopic cross section for pure UO_2 first. Next, we'll calculate the macroscopic cross section for pure graphite and then add that to the macroscopic cross section for UO_2 weighting both by their respective volume fractions.

Using the equations from above for "Calculating Macroscopic Cross Sections" for the case when atom fractions are given, $\Sigma = \rho \dfrac{N_a}{A} \sum\limits_{i=1}^{n} \gamma_i \sigma_i$, where $A = \sum\limits_{i=1}^{n} \gamma_i A_i$.

Given data: $\sigma_{t235} = 700$ b, $\sigma_{t238} = 12.2$ b, $\sigma_O = 4.03$ b, $\sigma_C = 4.74$ b, $\sigma_{UO2} = 11.0$ g/cm³, and $\sigma_C = 1.6$ g/cm³

First, we calculate the gram molecular weight A of pure UO_2.

$$A_{UO2} = (1)A_U + (2)A_O$$

Breaking down the gram atomic weight of uranium into its constituents,

$$A_{UO2} = (1)(0.08A_{235} + 0.92A_{238}) + (2)A_O$$

$$A_{UO2} = (0.08)(235 \text{ g/mol}) + (0.92)(238 \text{ g/mol}) + (2)(16 \text{ g/mol}) = 269.8 \text{ g/mol}.$$

Now applying the equation for calculating the macroscopic cross section in a similar fashion,

$$\Sigma_{UO2} = \left(11.0 \text{ g/cm}^3\right)(0.6022 \times 10^{24} \text{at/mol}) / \left(269.8 \text{ g/mol}\right)\left(\sum_{i=1}^{n} \gamma_i \sigma_i\right).$$

$$\Sigma_{UO2} = (2.455 \times 10^{22} \text{cm}^{-3})\left[(0.08)(700 \text{ b}) + (0.92)(12.2 \text{ b}) + (2)(4.03 \text{ b})\right]\left(10^{-24} \text{cm}^2/\text{b}\right)$$

$$\Sigma_{UO2} = (2.455'10^{22} \text{cm}^{-3})(75.28 \text{ b})\left(10^{-24} \text{cm}^2/\text{b}\right) = 1.848 \text{ cm}^{-1}.$$

Important: This is the total macroscopic cross section for a UO_2 fuel pellet, but not UO_2 mixed with graphite.
  The total macroscopic cross section for pure graphite is more easily calculated.

$$\Sigma_C = N_C \sigma_C = [\rho_{can} N_a / A_C] \sigma_C$$

$$\Sigma_C = [(1.6 \text{ g/cm}^3)(0.6022 \times 10^{24} \text{at/mol}) / (12 \text{ g/mol})](4.74 \text{ b})(10^{-24} \text{cm}^2/\text{b})$$

$$= 0.3806 \text{ cm}^{-1}$$

We are given the volume ratios of UO_2 and C, so the following equations apply:

$$\Sigma = \sum_{i=1}^{n} vf_i \Sigma_i,$$

where we have just calculated the Σ_i terms for pure substances, that is, UO_2 and C.

$$\Sigma_i = \rho_i \frac{N_a}{A_i} \sigma_i$$

Since UO_2 and C are mixed in a 1:3 ratio by volume, the volume fractions are $vf_{UO2} = 0.25$ and $vf_C = 0.75$; that is, $vf_{UO2}/vf_C = 1/3$.
  Therefore,

$$\Sigma_t = (0.25)\Sigma_{UO2} + (0.75)\Sigma_C$$

$$\Sigma_t = (0.25)\left(1.848 \text{ cm}^{-1}\right) + (0.75)\left(0.3806 \text{ cm}^{-1}\right) = 0.747 \text{ cm}^{-1}.$$

Why does it make sense to weigh the macroscopic cross sections of the two constituents by the volume fractions? We must remember that:
  Macroscopic cross section = (nuclide number density) × (microscopic cross section).

Weighing by the respective volume fractions is akin to multiplying the number densities by the respective volume which yields effectively the number of nuclei for each constituent.

8.10   PHOTON INTERACTIONS

The energy of a photon of frequency v: $E = hv$, where h is Planck's constant. Photon energies between 10 MeV and 20 MeV are important in radiation shielding design. In this range, the significant types of interactions are:

1. Photoelectric effect—predominates at lower energies.
2. Compton scattering (effect)—predominates at intermediate energies.
3. Pair production—important only for higher-energy photons.

8.11   PHOTOELECTRIC EFFECT

Photons incident on certain metals can result in electrons being ejected of the material's atoms. The photon energy is converted almost completely to the kinetic energy of an orbital electron. The **photoelectric effect** is dominant at low-energy photons. For light nuclei, all photoelectrons are typically K-shell electrons. For heavy nuclei, it will be about 80% from the K-shell electrons; approximation that total photoelectric cross section for heavy nuclei is 1.25 times cross section for K-shell electrons. The resulting electron vacancy is filled by an outer shell electron resulting in either the emission of fluorescence X-rays or Auger electrons. **Auger electrons** result when the hole in the K-shell is filled by an electron from an outer shell with the remaining energy transferred to another electron that is subsequently ejected. Energies of the fluorescence X-rays are unique for each element and are referred to as **characteristic X-rays**. If hv is photon energy and A is the binding energy of electron, known as the *work function*, then an energy of (hv - A) is distributed between photoelectron and recoil atom. Since $M_{atom} \gg m_e$, virtually all of the energy becomes the kinetic energy of the electron. The maximum energy of the photoelectron can be represented by $E_{max} = hv - A$. In the energy region where the photoelectric effect is dominant, $\sigma_{ph}(E) \propto Z^4/E^3$. Figure 8.14 shows an illustration of this effect.

Example 8.9

What is the maximum wavelength of light required to liberate photoelectrons from a metallic surface with a work function of 2.35 eV?

Solution

$$E = 0 = hv - A \rightarrow hv = A$$

$$v = A/h = (2.35 \, eV)/(4.136 \times 10^{-15} \, eV\text{-}s) = 5.68 \times 10^{-14} \, s^{-1}$$

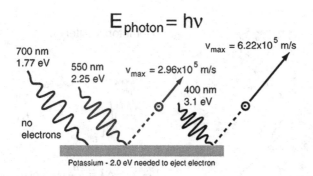

$E_{photon} = h\nu$

700 nm
1.77 eV

550 nm
2.25 eV

$v_{max} = 2.96 \times 10^5$ m/s

$v_{max} = 6.22 \times 10^5$ m/s

400 nm
3.1 eV

no
electrons

Potassium - 2.0 eV needed to eject electron

FIGURE 8.14 THE PHOTOELECTRIC EFFECT.

$$\lambda = c/\nu = (3.0 \times 10^8 \text{m/s})/(5.68 \times 10^{-14}/\text{s})$$

$$= 5.28 \times 10^{-7} \text{ m} = 5280 \text{ angstroms (green light)}$$

Note: 1 angstrom $= 10^{-8}$ cm $\approx$ diameter of electron field of an atom.

8.12   COMPTON SCATTERING

Based on the classical wave model of electromagnetic radiation, radiation should scatter from electrons with no change in the wavelength. In contraction to the classical model, in 1922, Arthur Compton observed X-rays scattering from electrons with a change in wavelength of $\Delta\lambda\alpha \cos(1-\theta)$. In scattering, the photon transfers a portion of its energy to an electron (see Figure 8.15).

Compton scattering is predominant for intermediate-energy photons. In actuality the interaction is with the entire atom with which the electron is bound. Although the interaction is inelastic, that is, kinetic energy is *not* conserved, because typical electron binding energy is typically significantly less than the photon energy, the interaction may be treated as elastic scatter of photon by a free electron.

The scattering of photons by atomic electrons may be classified as either ***incoherent* scattering** or *coherent* scattering. In incoherent scattering, the Compton equation for the free electron assumption breaks down as the kinetic energy of the recoil electron is comparable to its binding energy in the atom. However, in the region of interest, the photoelectric interaction cross sections are much greater than coherent scattering cross sections, so the latter are typically ignored.

In **coherent or Rayleigh scattering**, the scattering interaction is with multiple electrons of an atom simultaneously. Since in this case the recoil momentum is taken up by the atom as a whole, the energy loss by the photon is slight and therefore is typically ignored.

Assume the photons act like particles with linear momentum $p = h/\lambda$ and energy $E = h\nu = pc$. From the conservation of momentum,

$$p_{\lambda i} = p_{\lambda f} + p_e,$$

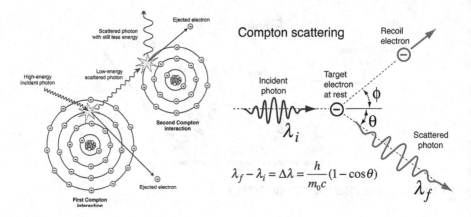

FIGURE 8.15 (Left) Multiple Compton scattering of a photon. (Right) Compton scattering schematic.

where p is the momentum vector and subscripts λi, λf, and e stand for incident photon, scattered photon, and recoiling electron, respectively. From the law of cosines,

$$p_e^2 = p_{\lambda i}^2 + p_{\lambda f}^2 - 2p_{\lambda i}p_{\lambda f}\cos\theta. \tag{8.1}$$

Assuming that the kinetic energy of the electron before collision is zero, from the conservation of energy:

$$p_{\lambda i}c + m_e c^2 = p_{\lambda f}c + mc^2$$

$$p_{\lambda i}c + m_e c^2 - p_{\lambda f}c = mc^2$$

$$p_{\lambda i} + m_e c - p_{\lambda f} = mc \text{ dividing through by c} \tag{8.2}$$

Aside: Recall from relativity theory that

$$m = \frac{m_e}{\sqrt{1 - \dfrac{v^2}{c^2}}} \rightarrow m^2\left(1 - \frac{v^2}{c^2}\right) = m^2 c^2 - m^2 v^2 = m_e^2 c^2$$

The momentum of the electron is $pe = mv$; thus, $m^2 c^2 = p_e^2 + m_e^2 c^2$ and $mc = [p_e^2 + m_e^2 c^2]^{1/2}$.

Substituting for *mc* into Eq. (8.2),

$$p_{\lambda i} + m_e c - p_{\lambda f} = [pe^2 + m_e^2 c^2]^{1/2} \tag{8.3}$$

Substituting for p_e^2 from Eq. (8.1) into Eq. (8.3) yields

$$p_{\lambda i} + m_e c - p_{\lambda f} = [p_{\lambda i}^2 + p_{\lambda f}^2 - 2p_{\lambda i}p_{\lambda f}\cos\theta + m_e^2 c^2]^{1/2}.$$

Squaring both sides and simplifying yields

$$2p_{\lambda i}m_e c - 2p_{\lambda i}p_{\lambda f} - 2p_{\lambda f}m_e c = -2p_{\lambda i}p_{\lambda f}\cos\theta.$$

Dividing through by $-2p_{\lambda i}p_{\lambda f}m_e c$ and rearranging gives

$$1/p_{\lambda f} - 1/p_{\lambda i} = (1 - \cos\theta)/(m_e c).$$

For a photon, $\lambda = h/p$

$$\lambda_f - \lambda_i = h(1 - \cos\theta)/(m_e c)$$

$$\lambda = c/\nu \text{ and } E = h\nu.$$

Therefore, $(1/E_f) - (1/E_i) = (1 - \cos\theta)/(m_e c^2)$, where E_i and E_f are the energies of the incident and scattered photon, respectively, and θ is the photon angle of scatter.

Example 8.10

What is the recoil kinetic energy of the electron that scatters a 3-MeV photon by 45 degrees?

Solution

We'll assume that KE of the electron before the collision was zero. Rearrange the above equation to find E_f

$$E_f = \left[\frac{1}{E_i} + \frac{(1-\cos\theta)}{(m_e c^2)}\right]^{-1}$$

$$E_f = \left[\frac{1}{3.0\,\text{MeV}} + \frac{1}{0.511\,\text{MeV}}(1-\cos(45°))\right]^{-1},$$

where $m_e c^2 = 0.511$ MeV.

$$E_f = [0.333 + 1.957(1 - 0.707)]{-1}\ \text{MeV} = 1.10\ \text{MeV}.$$

Because energy is conserved, the kinetic energy E_e of the recoil electron must equal the energy lost by the photon.

$$E_e = E_i - E_f = 3.0\ \text{MeV} - 1.10\ \text{MeV} = \textbf{1.90 MeV}.$$

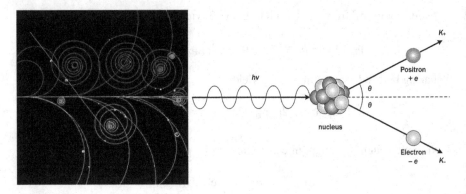

FIGURE 8.16 (Left) Electron and positron paths as a result of pair production and (Right) pair production diagram.

8.13 PAIR PRODUCTION

Recall from earlier in the chapter that a photon with a minimum energy of twice the rest mass energy of an electron can be transformed into an electron and positron. The following material is a review (Figure 8.16).

Pair production and pair annihilation are the quintessential examples of the conversion of energy into matter and vice versa. In pair production, a photon's energy is converted to mass and kinetic energy of an electron–positron pair. For this event to occur, the minimum energy of the photon must equal the combined rest-mass energies of the created electron and positron. Since the rest-mass energy of a positron or electron is 0.511 MeV, the photon energy threshold is greater than 1.022 MeV (i.e., 2×0.511 MeV); hence, predominant for high-energy photons. Due to the conservation of momentum after the pair production occurs, any excess energy of the original photon results in kinetic energy that is equally distributed between the positron and electron. The positron will eventually interact with an electron resulting in *pair annihilation* (see below).

$$E_+ + E_- = E_\gamma - 2m_e c^2,$$

where E_+ is the kinetic energy of the created positron, E_- is the kinetic energy of the created electron, E_γ is the energy of the photon, and $m_e c^2$ is the rest mass energy of an electron (or positron), which is 0.511 MeV. Thus, $\sigma_{pp}(E) \propto Z^2$. We can observe that $\sigma_{pp}(E)$ approaches a constant value at high-photon energy.

Example 8.11

A 5.022 MeV photon causes pair production. What are the resultant kinetic energies of the positron and electron?

Solution

Since the kinetic energy is equally distributed between the positron and electron, calculating the kinetic energy for one gives the value for the other.

$$2E_+ = E_\gamma - 2m_ec^2 = 5.022 \text{ MeV} - 2 \times 0.511 \text{ MeV} = 4 \text{ MeV}$$

Thus, $E_+ = 2.0$ MeV.

8.14 PAIR ANNIHILATION

The positron created in pair production will eventually interact with an electron resulting in two annihilation gammas, each with at a minimum energy of 0.511 MeV. This is for the limiting case where both the positron and electron possess zero kinetic energy at the time of their interaction. Energetic positrons will readily slow down in matter to low energy; thereafter, they are destined to form the metastable, pseudo-atom positronium by forming a mutually rotating, two-body, hydrogen-like object with an electron. This entity has an extremely short half-life ($\sim 10^{-10}$s) and decays by two quantum annihilation. Since the reactants possess little kinetic energy at annihilation, the two "annihilation photons" are of energy ~ 0.511 MeV. The net effect is conversion of one high-energy photon into two 0.511 MeV photons plus heat.

8.15 PHOTON ATTENUATION COEFFICIENTS

μ, the **photon linear attenuation coefficient**, in the limit of small paths lengths, is the probability per unit distance of travel that a photon undergoes any significant interaction. In other words, it is analogous to the macroscopic cross section for neutrons.

For photon energies of typical interest in nuclear engineering, the photon linear attenuation coefficient is the sum of components due to the photoelectric effect (subscript ph), Compton scatter (subscript c), and pair production (subscript pp). So,

$$\mu(E) = N[\mu_{ph}(E) + \mu_{inc}(E) + \mu_{pp}(E)] \approx N[\mu_{ph}(E) + \mu_c(E) + \mu_{pp}(E)],$$

where the material atom number density $N = \rho N_A/A$. It should be noted that Rayleigh coherent scattering and other minor effects were being omitted. Although μ_i is usually termed the *linear attenuation coefficient*, it is perhaps more appropriate to use the words **total linear interaction coefficient** since many interactions such as scattering do not "attenuate" the particle in the sense of an absorption interaction. Data for photon cross sections are typically provided not as linear attenuation coefficients but as **mass coefficients**, μ/ρ (typical reporting unit is cm^2/g), that is the value of the linear attenuation coefficient divided by the mass density of the material.

$$(\mu/\rho) = (N_A/A)[\mu_{ph}(E) + \mu_{inc}(E) + \mu_{pp}(E)] = \mu_{ph}/\rho + \mu_c/\rho + \mu_{pp}/\rho$$

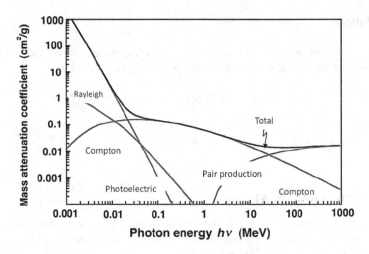

FIGURE 8.17 Variation of mass attenuation coefficients with photon energy.

So, it is independent of the density of the medium. For a given material table in Appendix, at low photon energies, μ_{ph}/ρ predominates, then with increasing energy μ_c/ρ, and finally with μ_{pp}/ρ (Figure 8.17).

8.16 SECONDARY RADIATION

Not all of an incident photon's energy is transferred to the local medium. Secondary photons (fluorescence, Compton scattered, annihilation, and Bremsstrahlung) are produced and continue to travel through the medium. If f is the fraction of the incident photon energy E that is transferred to secondary charged particles and subsequently transferred to the medium through ionization and excitation interactions,

$$\mu_{en} \approx \mu_f,$$

where μ_{en} is known as the linear energy absorption coefficient.

Similarly μ_{tr} is the *linear energy transfer coefficient* and is typically $\approx \mu_{en}$. Since like neutrons, photons can be considered "neutral" particles, the exact same concepts apply to photon attenuation as to neutron attenuation except for a change in terminology; that is, the neutron macroscopic cross section for reaction of type i, Σ_i, is analogous to the photon **linear coefficient** for reaction of type i, μ_i.

Example 8.12

In general, we can assume the compositions of air to be 75.3% nitrogen, 23.2% oxygen, and 1.4% argon by mass. Use the given data to determine the linear interaction coefficients in pure air at 20°C and 1 atm pressure for a 1-MeV photon and a thermal neutron (2200 m/s). Density of air at the given temperature and pressure is 0.001205 g/cm³.

Element	Photon	Neutron
	μ/ρ (cm²/g)	σ_{tot} (b)
Nitrogen	0.0636	11.9
Oxygen	0.0636	4.2
Argon	0.0574	2.2

Solution

Hence, density of each element is the density of air multiplied by the weight (mass) fraction for the element.

$$\mu = \rho_N(\mu/\rho)_N + \rho_O(\mu/\rho)_O + \rho_{Ar}(\mu/\rho)_{Ar} = [w_N(\mu/\rho)_N + w_O(\mu/\rho)_O + w_{Ar}(\mu/\rho)_{Ar}]\,\rho_{air}$$

$$\mu = [(0.753)(0.0636\,cm^2/g) + (0.232)(0.0636\,cm^2/g) + (0.014)(0.0574\,cm^2/g)]$$

$$\times\,(1.205 \times 10^{-3}\,g/cm^3) = \mathbf{7.65 \times 10^{-5}\,cm^{-1}}.$$

From these given data sets, it is possible to calculate the total macroscopic cross section.

$$\Sigma_t = N_N\sigma_N + N_O\sigma_O + N_{Ar}\sigma_{Ar}$$

where, for element i, $N_i = w_{i\,pair}N_a/A_i$.

$$\Sigma_t = [(w_N\sigma_N/A_N) + (w_O\sigma_O/A_O) + (w_{Ar}\sigma_{Ar}/A_{Ar})]\,\rho_{air}N_a$$

$$\Sigma t = [(0.753)(11.9\,b)/(14.0\,g/mol) + (0.232)(4.2\,b)/(16.0\,g/mol)$$

$$+ (0.014)(2.2\,b)/(39.9\,g/mol)]$$

$$\times\,(10^{-24}\,cm^2/b)(1.205 \times 10^{-3}\,g/cm^3)(0.6022 \times 10^{24}\,g/mol)$$

$$\Sigma_t = [\,0.640 + 0.061 + 0.001](7.26 \times 10^{-4}\,cm^{-1}) = \mathbf{5.10 \times 10^{-4}\,cm^{-1}}.$$

Here, we can see that at the typical energies of interest in nuclear engineering, photons in the form of gamma-rays or X-rays interact only with atomic electrons, whereas neutrons interact with the nuclei of atoms. Hence, the values of photon mass attenuation coefficients are identical for all isotopes of an element, whereas neutron microscopic cross sections vary from isotope to isotope of the same element.

8.17 RADIATION FLUX AND CURRENT

In defining the microscopic cross section, we introduced the idea of the "intensity" of a beam of particles. More specifically, we considered a group of neutrons, all of the same energy, that is, monoenergetic, moving parallel to each other. The intensity of the beam I represents the measured amount of neutrons that passed through a unit cross-sectional area perpendicular to the beam direction per unit second. Therefore, the beam intensity had unit of cm^{-2}-s^{-1}.

In the more general case, particles are not moving parallel to one another but are moving in many different directions. One measure of such a radiation field would be the particle number density at position r, that is, n(r). This is analogous to the atom number density N, but for particles. A more common and more useful measure to quantify the strength of a radiation field is the concept of **radiation or particle flux density**, more commonly referred to as simply the radiation or **particle flux**. Here, we also include the particle speed v in the definition, which in turn provides an indication of the energy of the particles.

Flux density is defined as

$$\phi(\mathbf{r}) = n(\mathbf{r})v, \text{ which is the unit of particles/cm}^2\text{-s,}$$

where **r** is the location vector at which the flux is measured. It is a scalar quantity. Flux can be thought of as the number of particles per unit time within a certain energy range entering a sphere with a unit cross section around point r, as shown in Figure 8.18.

Should the radiation of interest be photons, for example, gamma-rays or X-rays, the speed used in the flux equation is simply the speed of light; that is, if n(r) is the photon number density at location r, the photon flux at that location is

$$\phi(r) = n(r)c.$$

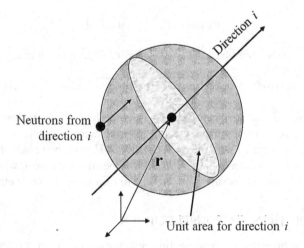

FIGURE 8.18 Illustration of the concept of radiation flux.

Although the precise definition of a radiation flux refers only to radiation of a specific speed or energy, in order to apply the concept in practical applications, it typically refers to radiation over a range of speeds or energies. This is for two reasons: (1) any scattering interaction a particle would result in a change to the particle's energy or speed, and (2) obviously speed is a continuous variable, so no two particles would ever have identical speeds and a radiation field would have to be described by an infinite number of fluxes. In the remainder of this course, we will typically ignore the energy dependence of flux, that is, the flux will refer to all particles of interest.

The radiation flux is a scalar quantity in that it accounts for all of the neutrons moving through a unit area regardless of direction. The *net* flow of neutrons in a specific direction is described by the *vector* J, termed **neutron current density vector** or simply neutron **current**. It has the same unit as radiation flux, that is, particles per square centimeter per second, but, being a vector quantity, also has a directional component. Unlike flux, current refers to radiation of a specific velocity and not speed, as velocity is a vector and speed a scalar. More practically speaking, current refers to the net radiation in a specific direction but over a range of speeds, that is, a range of velocity magnitudes.

The sum or integral of the radiation current at a specific location in a material over all directions is equal to the flux at that location. Now we can return back to our simple example of a beam of particles moving parallel to each other and perpendicularly incident on an infinite slab of material; the magnitude of the beam intensity I_0 in this case is equal to the magnitude of the particle current in the positive x-direction (refer back to Figure 8.10 (Left)); that is, now $I_0 = J_{x+}$.

For this example, we will ignore the energy dependence of the flux and current. Note that since some of the radiation that undergoes scattering collisions within the slab exits back through the left surface of the slab, that is, *backscatters*, the radiation flux ϕ_0 at the left slab surface is not equal to the beam intensity I_0 since the flux also includes this backscattered radiation. Similarly, the magnitude of the net radiation current J_0 in some direction at the left slab surface is not equal to the beam intensity I_0 at the surface. Based on Figure 8.12, the radiation flux at the left surface of the infinite slab in a vacuum includes all of the radiation incident on the slab from the left (all arrows) plus any backscattered radiation passing through the surface (green arrow). Deeper understanding of the difference between flux and current can be obtained by considering the problem of a small cubic volume being bombarded by perpendicularly incident beams of radiation on four of its surfaces as illustrated by Figure 8.19.

In this example, the total flux experienced by the cube of material is $I_1 + I_2 + I_3 + I_4$.

The net current in the positive x-direction is $J_x = (I_1 - I_3)I$, where i is the unit vector in the positive x-direction. Similarly, the net current in the positive y-direction is $J_y = (I_2 - I_4)j$, where j is the unit vector in the positive y-direction. It is important to remember that both radiation flux and current values apply to radiation over a certain speed range (or energy range). We will have frequent recourse to use of the concept of radiation flux, but not current, throughout the remainder of the textbook. We have introduced the concept of radiation current here as it becomes important in reactor theory.

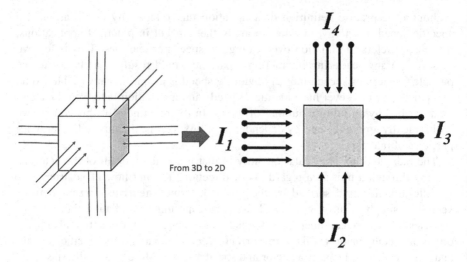

FIGURE 8.19 A material cube bombarded by four beams of radiation, I_1, I_2, I_3, and I_4.

8.18 RADIATION FROM AN ISOTROPIC POINT SOURCE IN A VACUUM

A **point source** is a mathematical idealization used to represent sources of radiation that are very small compared to other problem dimensions, as shown in Figure 8.20.

The magnitude of a point source, S_p, is given in the number of radiation particles emitted per unit time. Radiation is emitted in all directions with equal probability from an *isotropic* point; that is, the radiation from an isotropic point source spreads out equally in all directions. As a result of this spreading out, the radiation *flux* weakens in direct proportion to $1/r^2$, where r is the distance to a point of measurement from the point source. This weakening of radiation due only to the distance from the source is termed **geometric attenuation** since it is a function entirely of the geometry of the problem. If S_p is the number of particles per second emitted by an *isotropic* point source located in a *vacuum*, then at a distance *r* from the source, the radiation flux from the source is

$$\phi(r) = S_p/(4\pi r^2),$$

where $(1/r^2)$ factor is called the geometric attenuation term. This weakening of radiation due to the distance from a source is termed geometric attenuation. The fact that the weakening is inversely proportional to the square of the distance from the source is termed an **inverse-square law**, which is similar to the fall-off of a gravitational force (see Figure 8.21). It is important to emphasize that the measured flux is given by the above equation only appropriate to use when the radiation travel through a *vacuum*. Since in a vacuum radiation undergoes no interactions, such a flux is termed an **uncollided flux**.

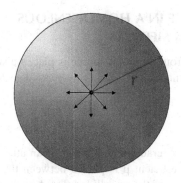

FIGURE 8.20 Point source of radiation.

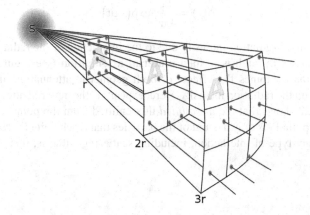

FIGURE 8.21 Illustration of "geometric attenuation." At a distance of 3r from the point source of radiation S, the same amount of radiation must disperse through an area nine times the area A traveled through at a distance of r from the point source.

Example 8.13

At a distance of 2 m from a point source in a vacuum, the radiation flux is measured to be 1600 cm^{-2}-s^{-1}. What is the measured radiation flux at distances of 4 m and 8 m from the source?

Solution

To answer this problem, we do not have to calculate the actual point source intensity, that is, S_p, although this could be done using the equation $\phi(r) = S_p/(4\pi r^2)$. Instead, we simply make use of the $1/r^2$ nature of point source flux in a vacuum, that is, $\phi_2 = (r_1/r_2)^2\phi_1$. So, $\phi_2(4\,m) = (2\ m/4\,m)^2(1600\,cm^{-2}\text{-}s^{-1}) = \mathbf{400\,cm^{-2}\text{-}s^{-1}}$ and $\phi_2(8\,m) = (2\ m/8\,m)^2(1600\,cm^{-2}\text{-}s^{-1}) = \mathbf{100\,cm^{-2}\text{-}s^{-1}}$.

It should be noted that in a vacuum, there is no material attenuation. Therefore, the *uncollided* flux is equal to the *actual* flux.

8.19 POINT SOURCE IN A HOMOGENEOUS ATTENUATING MEDIUM

Assume at a distance r from the point source is placed a small homogeneous mass with a volume ΔV_d. The interaction rate in this mass (detector) is

$$R(r) = \frac{\mu_d(E)\Delta V_d S_p}{(4\pi r^2)},$$

where $\mu_d(E)$ is the linear interaction coefficient. If an attenuating material of thickness t and attenuation coefficient μ is placed between the source and the point of measurement, the intensity of the uncollided flux ϕ at the point of measurement is then

$$\phi(r) = \frac{S_p}{4\pi r^2}\exp(-\mu t).$$

The weakening of radiation intensity due to the interaction of the radiation with the material it passes through is termed **material attenuation** (see Figure 8.22). The above equation combines the effects of both geometric attenuation and **material attenuation** on the radiation intensity. Here, exp(-μt) is the material attenuation *term*.

Figure 8.22 shows that the *uncollided* flux emitted from the point source S_p and measured at point P_1 is only a count of the particles that reach point P_1 *without* having undergone any type of interaction, including scattering—that is, their trajectory is

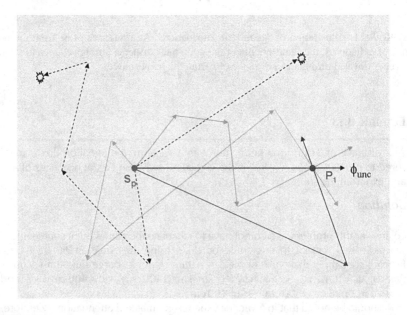

FIGURE 8.22 Point source embedded in a material.

represented by the black line since the path traveled from the point source to point P_1 is a straight line. Other particles emitted from the point source which might also happen to pass through point P_1 due to scattering (those indicated by the green, sky blue, and brown lines) are excluded from being counted as contributing to the *uncollided* flux. Such particles would be counted when measuring the *actual* flux at point P_1.

Example 8.14

A radioactive point source located in a vacuum with an activity of 1.86×10^{13} decays/s emits 2-MeV photons with a frequency of 70% per decay. What is the flux of 2-MeV photons 1 m from the source?

Solution

Since the medium is a vacuum, we ignore material attenuation. Therefore,

$$\phi(r) = S_p/(4\pi r^2) = (1.86 \times 10^{13} \text{ decays/s})(0.7 \text{ photos/decay})/[4\pi(100 \text{ cm})^2]$$

$$= 1.04 \times 10^8/\text{cm}^2\text{-s}.$$

Example 8.15

Based on the previous example, what thickness of iron needs to be placed between the source and a detector to reduce the *uncollided* flux to $2 \times 10^3/\text{cm}^2$-s? Use the fact that $(\mu/\rho)_{Fe} = 0.04254 \text{ cm}^2/\text{g}$ for 2-MeV photons and $\rho_{Fe} = 7.874 \text{ g/cm}^3$.

Solution

Here, $\phi/\phi_0 = \exp(-\mu_{Fe}x)$ where x equals the shield thickness.

$\phi_0 = 1.04 \times 10^8/\text{cm}^2$-s from the first part of the problem. Therefore,

$$\phi/\phi_0 = (2 \times 10^3/\text{cm}^2\text{-s})/(1.04 \times 10^8 \text{ cm}^2\text{-s}) = 1.96\text{v}10^{-5} = \exp(-\mu_{Fe}x)$$

$$\mu_{Fe} = (\mu/\rho)_{Fe} \, \rho_{Fe} = (0.04254 \text{ cm}^2/\text{g})(7.874 \text{ g/cm}^3) = 0.335 \text{ cm}^{-1}$$

$$x = -(1/\mu_{Fe}) \ln (\phi/\phi_0) = -(1/0.335 \text{ cm}^{-1}) \ln (1.96 \times 10^{-5}) = 32.4 \text{ cm}.$$

Example 8.16

Based on the first example, what is the thickness of a water shield and a lead shield needed to reduce a normally incident beam of 1-MeV photons to one-tenth of the incident intensity?

Given data: For water, $\mu/\rho(1 \text{ MeV}) = 0.07066 \text{ cm}^2/\text{g}$ and density is 1 g/cm^3; for Pb, $\mu/\rho(1 \text{ MeV}) = 0.06803 \text{ cm}^2/\text{g}$ and density is 11.35 g/cm^3.

Solution

For water, $\mu(1 \text{ MeV}) = (0.07066 \text{ cm}^2/\text{g})(1 \text{ g/cm}^3) = 0.07066 \text{ cm}^{-1}$. For Pb, $\mu(1 \text{ MeV}) = (0.06803 \text{ cm}^2/\text{g})(11.35 \text{ g/cm}^3) = 0.7721 \text{ cm}^{-1}$. Now,

$$\phi(x)/\phi_0 = \exp(-\mu_t x) \rightarrow \phi(x)/\phi_0 = 0.1 = \exp(-\mu_t x) \rightarrow \ln(0.1) = -2.30 = -\mu_t x.$$

So, $x = 2.30/\mu_t$

Water: $x = 2.30/(0.07066 \text{ cm}^{-1}) = \textbf{32.6 cm}$ and Pb: $x = 2.30/(0.7721 \text{ cm}^{-1}) = \textbf{2.98 cm.}$

8.20   SCATTERED RADIATION

In the above example, we calculated the thickness required to reduce the *uncollided* beam by 90%. At any distance x into an attenuating material, the population of radiation particles consists of both uncollided and scattered (collided) particles. The calculation of the number of particles that have undergone one or more scattering collisions requires the use of *particle transport* theory which is beyond the scope of this course.

In the case of photon radiation, the total radiation field (collided plus uncollided) can be obtained from the uncollided beam intensity as follows:

$$\phi(x) = B(x) \, \phi_{\text{uncollided}}(x),$$

where B(x) is termed the **buildup factor**. Values of build factors are calculated for various conditions from particle transport codes and made available in tabular form.

Actual flux at any point in a material medium will always exceed the uncollided flux. This is shown in the following plot of actual versus uncollided neutron flux through a slab with an incident monoenergetic neutron beam (refer back to Figure 8.11). Practical problems often require calculating the *total interaction coefficients* (*macroscopic cross sections*) for a material composed of two or more types of nuclides.

Example 8.17

What is the total interaction coefficient (μ_t) for a 1-MeV photon in a medium composed of a homogeneous mixture of iron and lead with equal proportion by weight?

Solution

$$(\mu/\rho)_{\text{mix}} = wf_{\text{Fe}}(\mu/\rho)_{\text{Fe}} + wf_{\text{Pb}}(\mu/\rho)_{\text{Pb}.}$$

Using values for 1-MeV photon from Appendix,

$$(\mu/\rho)_{mix} = (0.5)(0.05951 \text{ cm}^2/\text{g}) + (0.5)(0.06803 \text{ cm}^2/\text{g}) = 0.06377 \text{ cm}^2/\text{g}.$$

Assuming that upon mixing there is no volume change (*not* necessarily true), then 2 g of the resultant mixture contains a gram each of Fe and Pb. Hence, 2 g of the material has a volume of $(1 \text{ g}/\rho_{Fe}) + (1 \text{ g}/\rho_{Pb}) = (1 \text{ g}/7.784 \text{ g/cm}^3) + (1 \text{ g}/11.35 \text{ g/cm}^3) = 0.2151 \text{ cm}^3$
The mixture density is thus: $\rho_{mix} = 2 \text{ g}/0.2151 \text{ cm}^3 = 9.298 \text{ g/cm}^3$.

$$\mu_{mix} = \rho_{mix}(\mu/\rho)_{mix} = (9.298 \text{ g/cm}^3)(0.06377 \text{ cm}^2/\text{g}) = \mathbf{0.593 \text{ cm}^{-1}}$$

8.21   UNCOLLIDED VERSUS ACTUAL FLUX

Herein we explore three cases, the goal of which is to provide a deeper understanding of the difference between the "uncollided" (or "virgin") radiation flux and the "actual" radiation flux. For the first case, consider an isotropic point source S_p of radiation embedded in a homogeneous material as shown in Figure 8.22. The problem is to determine the radiation flux from the point source that would be measured at point P_1. The *uncollided* flux at point P_1 is a measure of those particles emitted from the point source that reach point P_1 without having undergone any scattering collisions, that is, those particles whose path is the straight line connecting S_p to point P_1 as shown by the solid black line.

The *actual* flux at point P_1 includes those particles making up the uncollided flux plus any other particles emitted from the point source whose path, after undergoing one or more scattering collisions, also passes through point P_1. In Figure 8.22, the paths of such particles are indicated by the green, sky blue, and brown lines. If the distance from the point source S_p to the point of measurement P_1 is r, and the total macroscopic cross section for the homogeneous medium is Σ_t, then the value of the *uncollided* flux at P_1 is given by

$$\varphi_{unc} = \frac{S_p e^{-\Sigma_t r}}{4\pi r^2}.$$

Here, the total attenuation effect on the radiation emitted from the point source is the product of the inverse-square geometric attenuation given by the $1/(4\pi r^2)$ multiplier and the material attenuation given by the $\exp(-\Sigma_t r)$ multiplier. The distance term r in both multipliers is the same since both geometric and material attenuation apply over the same distance. In order to calculate the *actual* particle flux measured at point P_1, we would need to use a more sophisticated means of calculation such as radiation transport theory or Monte Carlo method.

The second case is identical to the first except the material medium is replaced by a vacuum, as shown in Figure 8.23. In this case, no material attenuation occurs. Therefore, the *actual* flux measured at point P_1 is identical to the *uncollided* flux since no particle scattering occurs. The flux is then simply given by

$$\varphi = \frac{S_p}{4\pi r^2}.$$

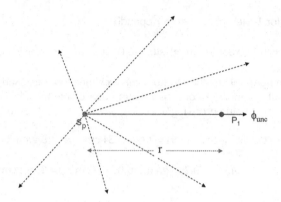

FIGURE 8.23 Isotropic point source S_p located in a vacuum.

For the third case, a homogeneous slab shield of thickness x is inserted between the point source and the point of measurement as shown in Figure 8.24. We are asked to determine the *uncollided* flux at point P_1. This type of problem often troubles the student in that the exact location of the shield is not specified. Essentially the slab is specified to be perpendicular to the line connecting the point source to P_1, that is, the path of the *uncollided* particles, and the entire shield is between the two points of interest, that is, neither the point source nor the point of measurement is located within the shielding material.

Students are confused in that they intuitively deduce that the value of the *actual* flux measured at point P_1 does depend on the location of the shield. That is indeed true. So if that is true, why would the value of the *uncollided* flux be independent of the shield location?

To better understand why the *uncollided* flux value depends only on the shield thickness and not on its location, consider the geometries depicted in Figures 8.24 and 8.25 (Left) compared to that of Figure 8.24. In Figure 8.25 (left), the left surface of the shield is aligned with the point source, and conversely, the other extreme of the case geometry is depicted in Figure 8.25 (right), where the right surface of the shield is aligned with point P_1.

In summary, the geometric attenuation of the beam at point P_1 is identical for all three cases since the distance between the source and point P_1 (which is r) is the same for all three cases. The path taken by the *uncollided* particles for all three cases is indicated by the black line. The distance over which the *uncollided* beam of particles experiences material attenuation, x, is identical for all three cases. Hence, the value of the *uncollided* flux at point P_1 is unchanged regardless of the location of the shield between the source and point of measurement. Unlike the value of the uncollided flux, the value of the *actual* flux at point P_1 will vary depending on shield location.

Thus, the value of the *uncollided* flux for the third case is given by

$$\varphi_{unc} = \frac{S_p e^{-\Sigma_t x}}{4\pi r^2}.$$

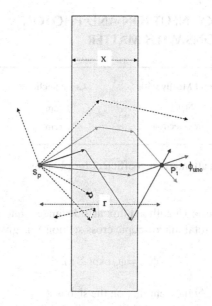

FIGURE 8.24 Shield attenuation case with the shield of thickness × located approximately mid-way between the point source and the point of measurement.

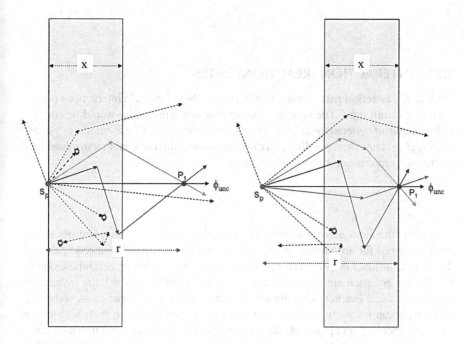

FIGURE 8.25 (Left) Shield attenuation case with the shield left-aligned with the point source. (Right) Shield attenuation case with the shield right-aligned with the point of measurement.

8.22 SUMMARY OF NEUTRON AND PHOTON INTERACTIONS WITH MATTER

Radiation	Interact Mostly with	Cross Section*	How Calculated
Neutrons	Nuclei	Σ (cm^{-1})	$\Sigma = N\sigma$
Photons	Atomic electrons	μ (cm^{-1})	$\mu = \rho(\mu/\rho)$

*Also called the linear attenuation coefficient for photons.

Infinite slab problem: Uncollided flux at a distance x into a homogeneous, infinite slab material with total macroscopic cross section Σ_t is given by

$$\phi_{unc} = \phi_0 \exp(-\Sigma_t x).$$

Here, the perpendicularly incident flux on the slab is ϕ_0.

Point source problem: In general, the uncollided flux, ϕ_{unc}, at a distance r from a point source S_p in a homogeneous material with total macroscopic cross section Σ_t is given by

$$\varphi_{unc} = \frac{S_p e^{-\Sigma_t r}}{4\pi r^2}.$$

8.23 INTERACTION (REACTION) RATES

Let R be the **reaction rate** density, that is, the number of particle interactions per unit volume and unit time. The common unit of reaction rate density would be cm^{-3}-s^{-1}. *The concept of reaction rate density is one of the utmost importance in nuclear engineering.* The corresponding equation for neutron interactions in a material characterized by microscopic cross section σ_i is

$$R_i(\text{cm}^{-3}\text{-s}^{-1}) = (N\sigma_i)(nv),$$

where N is the atom number density of material, σ_i is the microscopic cross section of the material for an interaction of type i for particles moving at speed v, and n is the particle number density. The two terms inside the first set of parentheses on the RHS of the equation are mostly a function of the material in which the interactions are taking place. For neutron interactions, this is the macroscopic cross section, Σ_i, and for photon interactions, the linear attenuation coefficient, μ, both with unit of cm^{-1}. In general, Σ_i and μ provide the probability of an interaction per unit path length of radiation in some homogeneous material. The two terms inside of the second set of parentheses on the RHS are usually combined into the radiation flux ϕ with unit of cm^{-2}-s^{-1}. The flux is characteristic of the particle population and energy.

In its more common form, the reaction rate density equation is stated as the product of a cross section (or linear attenuation coefficient) and flux; that is,

$$R_i(r) = \Sigma_i(r)\phi(r) \text{ for neutrons AND } R_i(r) = \mu_i(r)\phi(r) \text{ for photons.}$$

Example 8.18

A sample of ^{10}B ($\rho = 2.3$ g/cm^3) is exposed to a thermal neutron flux of 10^{13}/cm^2-s. What is the *absorption* reaction rate density and *total* reaction rate density of the sample?

Given data: Thermal neutron cross sections for ^{10}B: $\sigma_\alpha = 3840$ b; (n, α) reaction: $\sigma_\gamma = 0.50$ b; (n.γ) reaction: $\sigma_t = 3847$ b total cross section.

Solution

Hence, the total microscopic absorption cross section is: $\sigma_a = \sigma_\alpha + \sigma_\gamma = 3840 + 0.5 = 3841$ b. We wish to find $R_t = \Sigma_t\phi$ and $R_a = \Sigma_a\phi$. In order to calculate the macroscopic cross sections, we first need to calculate the boron-10 atom number density (N), $N = (\rho N_a/A)$ since $\Sigma_i = N\sigma_i$.

$$N = (2.3 \text{ g/cm}^3)(0.6022 \times 10^{24} \text{ at/mol})/(10 \text{ g/mol}) = 0.139 \times 10^{24} \text{ at/cm}^3$$

$$\Sigma_a = N\sigma_a = (0.139 \times 10^{24} \text{ at/cm}^3)(3841 \text{ b})(10^{-24} \text{cm}^2/\text{b}) = 534 \text{cm}^{-1}$$

$$\Sigma_t = N\sigma_t = (0.139 \times 10^{24} \text{ at/cm}^3)(3847 \text{ b})(10^{-24} \text{cm}^2/\text{b}) = 538 \text{cm}^{-1}$$

$$R_a = \sigma_a\phi = (534 \text{cm}^{-1})(10^{13}/\text{cm}^2\text{-s}) = 5.34 \times 10^{15}/\text{cm}^3\text{-s}$$

$$R_t = \sigma_t\phi = (538 \text{cm}^{-1})(10^{13}/\text{cm}^2\text{-s}) = 5.38 \times 10^{15}/\text{cm}^3\text{-s}.$$

The difference between R_t and R_a is the scattering reaction rate density.

Example 8.19

Find the average neutron density n corresponding to a typical LWR flux ϕ of 5×10^{13} cm^{-2}-s^{-1}, assuming an effective neutron speed v of 2200 m/s.

Solution

$\phi = nv \rightarrow n = \phi/v$. So, $n = 5 \times 10^{13}$ cm^{-2}-s^{-1}/ 2.2×10^5 cm/s $= 2.275 \times 10^8$ cm^{-3}.

Example 8.20

For equal nuclide densities of ^{235}U and ^{239}Pu in a given reactor, find (a) the fraction of fissions for each and (b) the absorption mean free path for each nuclide and for the mixture. Assume each has an atom density of 10^{21} cm^{-3}.

Solution

a. Fission fraction: $^jF = (^j\Sigma_{f\phi})/(^{mix}\Sigma_{f\phi}) = (^jN^j\sigma_f)/[^{235}N^{235}\sigma_f + {^{239}}N^{239}\sigma_f]$
   Since $^{235}N = {^{239}}N$, $^jF = {^j}\sigma_f/(^{235}\sigma_f + {^{239}}\sigma_f)$ and $^{235}F = (585\text{ b})/(585\text{ b} + 750\text{ b}) = \mathbf{0.44}$
   and $^{239}F = 1 - {^{25}}F = \mathbf{0.56}$.

b. Mean free paths:

$$^{235}\Sigma_a = {^{235}}N(^{235}\sigma_\gamma + {^{239}}\sigma_f) = (10^{21}\text{ at/cm}^3)(10^{-24}\text{ cm}^2/\text{b})(99\text{ b} + 585\text{ b}) = 0.684\text{ cm}^{-1}$$

$$^{235}\Sigma_a = (10^{21}\text{ at/cm}^3)(10^{-24}\text{ cm}^2/\text{b})(271\text{ b} + 750\text{ b}) = 1.02\text{ cm}^{-1}$$

$$^{mix}\Sigma_a = {^{235}}\Sigma_a + {^{239}}\Sigma_a = 1.704\text{ cm}^{-1}$$

$$^{235}\lambda_a = 1/^{235}\Sigma_a = \mathbf{1.46\,cm},\ ^{239}\lambda_a = 1/^{239}\Sigma_a = \mathbf{0.979\,cm},\ ^{mix}\lambda_a = 1/^{mix}\Sigma_a = \mathbf{0.587\,cm}$$

8.24   RADIATION FLUENCE

In many situations, the total exposure to radiation over some period of time is required. This is typically given as the fluence Φ, that is, the integral of the radiation flux over a period of time, yielding unit of cm^{-2}, which is

$$\Phi(\mathbf{r},\ E) \equiv \int_{t1}^{t2} \phi(\mathbf{r},\ E,\ t)\,dt.$$

8.25   BREMSSTRAHLUNG RADIATION

Neutral particles such as neutrons and photons have no definite range and are exponentially attenuated as they traverse a medium. In contrast, charged particles continuously interact with the atomic electrons of the ambient atoms resulting in excitation and ionization of the surrounding medium. Thousands of such interactions are typically required to transfer the particle's kinetic energy to the medium and slow the particle to thermal energies.

As with neutral particles, charged particle may on occasion cause a nuclear reaction or transmutation, or scatter from a nucleus. If the energy of a charged particle sufficiently high (or mass sufficiently small), the particle will lose energy through the emission of electromagnetic radiation as they are deflected (centripetally accelerated) by atomic nuclei. High-energy X-rays typically produced by this phenomenon are termed **Bremsstrahlung** radiation.

8.26   PARTICLE RANGE

The paths traveled by light charged particles (beta) vary tremendously due to the possibility of large-angle scattering which in turn produces secondary electrons (also called *delta* rays) with substantial recoil energy. The range of beta particles is defined

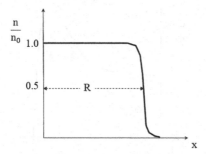

FIGURE 8.26 Plot of fast ion attenuation in a thick absorber. Incident ion starts off at the left side in a thick absorber, continuous lose energy as they traverse in the medium. There is little loss of particles until R.

as the total path length—although herein we shall define light particle range as *mean free path*.

Heavy charged particles (protons and up) move in almost straight lines since these particles typically experience negligible deflection due to the large mass ratio between the heavy charged particle and an atomic electron. For example, consider the alpha particle. At the most, the maximum energy loss of a 4-MeV alpha interacting with an electron is 2.2 keV, and typically, significantly less. Hence, it requires thousands of interactions for an alpha particle to lose its kinetic energy. The **continuous slowing-down approximation**, R, is a standard means of specifying the penetration depth of a heavy charged particle (see Figure 8.26).

8.27 STOPPING POWER

A charged particle loses energy as a result of: (1) Coulombic interactions with (atomic) electrons—predominate mechanism for heavy charged particles and (2) Radiation losses (bremsstrahlung). Only for very low-energy particles are collisions with atomic nuclei important. Here, we can define:

Collisional stopping power (ionization stopping power): result of Coulombic interactions, $(-dE/ds)_{coll}$

Radiative stopping power: $(-dE/ds)_{rad}$.

Now, R can be represented by

$$R = \int_0^R ds = \int_{E0}^0 \frac{dE}{(-dE/ds)_{tot}},$$

where $(-dE/ds)_{tot} = (-dE/ds)_{coll} + (-dE/ds)_{rad}$.

Figure 8.27 shows that the gradual increase in $-dE/dx$ is indicative of the $1/E$ dependence in the stopping power. Charge neutralization and eventual stopping are responsible for the sharp drop in $-dE/dx$ at the end of the particle trajectory.

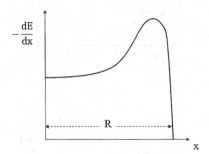

FIGURE 8.27 Behavior of the stopping power expression along charged projectile path in a thick absorber for heavy charged particles incident from the left at x = 0.

8.28 IONIZING RADIATION FOR α, β, AND FISSION FRAGMENTS

Let LET represents *linear energy transfer*, which is LET = dE/dx and R be the range—the average distance a charged particle travels before it is stopped; that is, the average distance traveled by a charged particle of initial energy E_0 until its energy is zero.

$$R = \int_0^{E_0} \frac{dE}{LET} = \int_0^{E_0} \frac{dE}{dE/dx}.$$

Based on the above equation, we can observe the following:

- i. Fission fragments: masses between about 80 and 150 amu, charge about +20e at the time of fission, R = few cm in air, fraction of mm in a solid.
- ii. α: R ≈ 3–6 times that of fission fragments; LET about 10^{-1} that of fission fragment.
- iii. β: R ≈ 100 times that of α; LET about 10^{-2} that of α.

8.29 ELECTROMAGNETIC RADIATION

Photons interact directly with electrons, more rarely with nuclei; hence, interactions depend more on the density of electrons than on the specific atom composition. As a result, the cross sections (linear attenuation coefficients) are relatively material-independent. In general, R ≈ 100 times that of β, and LET is about 10^{-2} that of β.

8.30 PRACTICE SEVERAL EXAMPLES

Example 8.21

Assume that the microscopic fission, absorption, and scattering cross sections for ^{235}U for thermal neutrons are 505 b, 591 b, and 15 b, respectively. What is the probability that if a neutron interacts with a ^{235}U nucleus, it will be absorbed? What is the probability that if a neutron is absorbed by a ^{235}U nucleus, it will cause fission? What is the probability of radiative capture of a neutron that is absorbed by a ^{235}U nucleus?

Solution

$$p_a = \sigma_a/\sigma_t = \sigma_a/(\sigma_a + \sigma_s) = 591 \text{ b}/(591 \text{ b} + 15 \text{ b}) = 0.975$$

$$p_f = \sigma_f/\sigma_a = 505 \text{ b} / (591 \text{ b}) = \mathbf{0.854}.$$

In ^{235}U as with most actinides, the only significant absorption mechanisms are fission and radiative capture. Therefore, an absorbed neutron that did not cause a fission must result in radiative capture, which is $p_c = 1 - p_f = 1 - 0.854 = \mathbf{0.146}$.

Example 8.22

A boiling water reactor operates at 1000 psi. At that pressure, the density of water and of steam are, respectively, 0.74 g/cm³ and 0.036 g/cm³. The microscopic *total* cross sections of H and O are 21.8 b and 3.8 b. (a) What is the macroscopic total cross section of the water? (b) What is the macroscopic total cross section of the steam? (c) If, on the average, 40% of the volume is occupied by steam, then what is the macroscopic total cross section of the steam–water mixture? (d) What is the macroscopic total cross section of water under atmospheric conditions at room temperature? *Note:* Assume that the microscopic cross sections are additive for water—in reality, at thermal neutron energies they are not.

Solution

a. $N_w = \rho_w N_A/A_w = (0.74 \text{ g/cm}^3)(0.6022 \times 10^{24} \text{ at/mol})/(18 \text{ g/mol}) = 0.0248 \times 10^{24}$ at/cm³ and $\Sigma_w = N_w(2\sigma_H + \sigma_O) = (0.0248 \times 10^{24} \text{ at/cm}^3)(2 \times 21.8 \times 10^{-24} \text{ cm}^2 + 3.8 \times 10^{-24} \text{ cm}^2) = \mathbf{1.18 \text{ cm}^{-1}}$.
b. For steam, the only difference in the above equation is the density.

$\Sigma_s = \Sigma_w (\rho_s/\rho_w) = (1.18 \text{ cm}^{-1})(0.036 \text{ g/cm}^2)/(0.74 \text{ g/cm}^3) = 0.0574 \text{ cm}^{-1}$
c. 40% steam mixture, $\Sigma_{mix} = (0.4) \Sigma_s + (0.6) \Sigma_w = (0.4)(0.0574 \text{ cm}^{-1}) + (0.6)(1.18 \text{ cm}^{-1}) = \mathbf{0.731 \text{ cm}^{-1}}$.
d. At STP, $\Sigma_{STP} = \Sigma_w (\rho_{STP}/\rho_w) = (1.18 \text{ cm}^{-1})(1.0 \text{ g/cm}^3)/(0.74 \text{ g/cm}^3) = \mathbf{1.59 \text{ cm}^{-1}}$.

Example 8.23

What is the total macroscopic thermal cross section of uranium dioxide (UO_2) that has been enriched to 4% (assume a/o)? Assume $\sigma_{235} = 607.5$ b, $\sigma_{238} = 11.8$ b, $\sigma_O = 3.8$ b, and that UO_2 has a density of 10.5 g/cm³.

Solution

Here, $\sigma_U = \gamma^{235}\sigma^{235} + \gamma^{238}\sigma^{238} = (0.04)(607.5 \text{ b}) + (0.96)(11.8 \text{ b}) = 35.6 \text{ b}$

$$\sigma_U O^2 = \gamma_{U\sigma U} + \gamma_{O\sigma O} = 35.6 \text{ b} + 2(3.8 \text{ b}) = 43.2 \text{ b}$$

$$\Sigma_U O^2 = \rho_w N_A \sigma_a / A$$

$$= (10.5 \text{ g/cm}^3)(0.6022 \times 10^{24}/\text{mol})(43.2 \times 10^{-24} \text{ b})/(238 \text{ g} + 32 \text{ g}) = \mathbf{1.01 \text{ cm}^{-1}}.$$

Example 8.24

Equal volumes of graphite and iron are mixed together. Fifteen percent of the volume of the resulting mixture is occupied by air pockets. Find the total macroscopic cross section given the following data: $\sigma_C = 4.75$ b, $\sigma_{Fe} = 10.9$ b, $\rho_C = 1.6$ g/cm³, $\rho_{Fe} = 7.7$ g/cm³. Is it reasonable to neglect the cross section of air? Why?

Solution

$\Sigma_t = V_{fC}\rho_C N_A \sigma_C / A_C + V_{fFe}\rho_{Fe} N_A \sigma_{Fe} / A_{Fe}$, where V_{fC} and V_{fFe} are the volume fractions of graphite and iron, respectively. So, $V_{fC} = V_{fFe} = (1 - 0.15)/2 = 0.425$, and

$\Sigma_t = V_{fC} N_A [\rho_C \sigma_C / A_C + \rho_{Fe} \sigma_{Fe} / A_{Fe}] = (0.425)(0.6022 \times 10^{24} /\text{mol})[(1.6 \text{ g/cm}^3)(4.75 \times 10^{-24} \text{cm}^2)(12 \text{ g/mol}) + (7.7 \text{ g/cm}^3)(10.9 \times 10^{-24}\text{cm}^2)(55.85 \text{ g/mol})] = \mathbf{0.547 \text{ cm}^{-1}}.$

The cross section of air can be neglected due to its extremely small atom density compared to graphite and iron.

Example 8.25

Assume that the microscopic absorption and scattering cross sections for ^{56}Fe for thermal neutrons are 2.29 b and 11.3 b, respectively. In a sample of this iron with a mass density of 7.8 g/cm³, what is the average distance (mean free path) a thermal neutron will travel before undergoing an interaction with an iron nucleus? What is the average distance such a neutron will travel before being absorbed by an iron nucleus?

Solution

The number density of the iron atoms is $N = \rho N_A / A$.

$$N = (7.8 \text{ g/cm}^3)(0.6022 \times 10^{24} \text{ at/mol})/(56 \text{ g/mol}) = 8.39 \times 10^{22} \text{ at/cm}^3.$$

The total mean free path can be calculated.

$$\lambda_t = 1/\Sigma_t = 1/(N\sigma_t) = 1/[(8.39 \times 10^{22} \text{ at/cm}^3)(2.29 \text{ b} + 11.3 \text{ b})(10^{-24}\text{cm}^2/\text{b})] = \mathbf{0.88 \text{ cm}}.$$

The absorption mean free path can be calculated.

$$\lambda_a = 1/\Sigma_a = 1/(N\sigma_a) = 1/[(8.39 \times 10^{22} \text{ at/cm}^3)(2.29 \text{ b})(10^{-24}\text{cm}^2/\text{b})] = \mathbf{5.2 \text{ cm}}.$$

TABLE 8.1
Neutron and Photon Interactions

Phenomenon	Neutrons	Photons
Characterizes the probability of an interaction with a single atom	σ—microscopic cross section	Not applicable
Probability per unit path length traveled that an interaction will occur	Σ—macroscopic cross section	μ—linear attenuation coefficient

8.31 SUMMARY

Over the range of most energies of interest in nuclear engineering: (1) neutrons interact with individual nuclei, and (2) photons interact with atomic electrons or undergo pair production (see Table 8.1).

If N is the atom number density of a material whose microscopic cross section is σ, then $\Sigma = N\sigma$. If ρ is the mass density of a material whose mass coefficient is (μ/ρ), then $\mu = \rho(\mu/\rho)$. Macroscopic cross sections Σ and linear attenuation coefficients μ are scalar quantities, that is, they are additive. The typical unit is cm^{-1}.

Next, $\phi = nv$, where n is the particle number density of the radiation and v is the particle speed (= c for photons). The typical unit is $cm^{-2}\text{-}s^{-1}$. A *beam* of radiation is radiation flux where the particle paths are parallel to one another, often denoted by I instead of ϕ. Radiation flux ϕ is a scalar quantity, that is, fluxes are additive. Summary of these equations and characterization are listed in Table 8.2.

Important Terms (in the Order of Appearance)

Directly ionizing radiation	Mean time	Neutron current density vector
Indirectly ionizing radiation	Photoelectric effect	Neutron current
Secondary radiation	Auger electrons	Point source
Barn	Characteristic X-rays	Geometric attenuation
Thermal energy	Compton scattering	Inverse-square law
1/v absorption	Incoherent scattering	Uncollided flux
Resonance absorption or resonance	Coherent or Rayleigh scattering	Material attenuation
Fast energy	Pair production	Buildup factor
Potential scattering	Pair annihilation	Reaction rate
Resonance scattering	Photon linear attenuation coefficient	Bremsstrahlung
Macroscopic cross section	Total linear interaction coefficient	Continuous slowing-down approximation
Attenuation	Mass coefficient	Stopping power
Mean free path	Flux	Collisional stopping power
Collision frequency	Flux density	Radiative stopping power

TABLE 8.2
Radiation Flux and Reaction Rate for Neutrons and Photons

Phenomenon	Neutrons	Photons
Uncollided radiation flux due to material attenuation of a radiation beam	$I_{unc}(x) = I_0 e^{-\Sigma x}$	$I_{unc}(x) = I_0 e^{-\mu x}$
Radiation flux from a point source in a vacuum	$\phi(x) = S_p \dfrac{1}{4\pi r^2}$	$\phi(r) = S_p \dfrac{1}{4\pi r^2}$
Uncollided radiation flux from a point source in a material	$\phi_{unc}(r) = S_p \dfrac{e^{-\Sigma r}}{4\pi r^2}$	$\phi_{unc}(r) = S_p \dfrac{e^{-\mu r}}{4\pi r^2}$
Reaction rate density, cm^{-3}-s^{-1}	$R = \Sigma\phi$	$R = \mu\phi$

BIBLIOGRAPHY

Bush, H.D., *Atomic and Nuclear Physics,* Prentice Hall, Englewood Cliffs, NJ, 1962.

Coombe, R.A., *An Introduction to Radioactivity for Engineers.* Macmillan Publishing Co., Inc., New York, 1968.

Glasstone, S., and R.H. Loveberg, *Controlled Thermonuclear Reactions,* D. Van Nostrand Company, Princeton, NJ, 1960.

Keepin, G.R., *Physics of Nuclear Kinetics,* Addison-Wesley, New York, 1965.

Lamarsh, J.R., and Baratta, A.J., *Introduction to Nuclear Engineering*, 3rd Ed., Prentice Hall, Englewood Cliffs, NJ, 2001.

Lewis, E.E., *Nuclear Reactor Physics,* Elsevier, New York, 2008.

Mayo, R.M., *Nuclear Concepts for Engineers*, American Nuclear Society, La Grange Park, IL, 1998.

Shultis, J.K. and R.E. Faw, *Fundamentals of Nuclear Science and Engineering* 2nd Ed., CRC Press, Taylor & Francis Group, Boca Raton, FL, 2008.

FURTHER EXERCISES

A. True or False: If the statement is false, give a counterexample or explain the correction. If the statement is true, explain why it is true.

1. Microscopic cross section is the probability that a particle will interact with a nucleus.
2. The total macroscopic cross section is the summation of the scattering and absorption macroscopic cross sections.
3. If the absorption cross section is 30 b and the capture cross section is 10 b, then the fission cross section is 40 b.
4. The linear absorption coefficient has the same units as the macroscopic cross section.
5. Fluence is a measure of strength or intensity of a radiation field.
6. Fluence is a measure of absorbed dose for radiation damage in materials.
7. Reaction rate is the product of flux, atom density, and microscopic cross section.

8. Flux for point source in vacuum followed the square law and is referred to as geometric attenuation.

9. Bremsstrahlung is one of the secondary radiations.

10. Attenuation coefficient is used for gamma while macroscopic cross section should be used for neutron.

B. Problems

1. A broad beam of neutrons is normally incident on a magic slab of 0.06 m thick. The intensity of neutrons transmitted through the slab *with* interaction is found to be 70% of the incident intensity. (a) What is the total *macroscopic cross section* for the slab material *in cm*$^{-1}$? (b) What is the average distance a neutron travels in the material in *cm* before undergoing an interaction?

2. In natural uranium, 0.720% of the atoms are the isotope ^{235}U, 0.0055% are ^{234}U, and the remainder ^{238}U. (a) What is the total macroscopic cross section in cm^{-1} for a thermal neutron in natural uranium? (b) What is the total macroscopic fission cross section in cm^{-1} for thermal neutrons? Take r of natural U $= 18.95$ g/cm^3.

3. Stainless steel, type 304 having a density of 7.86 g/cm^3, has been used in some reactors. The nominal composition by weight of this material is as follows: carbon, 0.08%; chromium, 19%; nickel, 10%; iron, the remainder. Calculate the macroscopic absorption cross section of SS-304 for 0.0253 eV neutrons assuming the following absorption cross sections: C, 0.0034 b; Cr, 3.1 b; Ni, 4.43 b; Fe, 2.55 b.

4. A 1-mCi source of ^{60}Co is placed in the center of a cylindrical water-filled tank with an inside diameter of 20 cm and depth of 100 cm. The tank is made of iron with a wall thickness of 1 cm. What is the *uncollided* flux density in particles per cm^2-s at the outer surface of the tank nearest the source? Making the right assumption(s) simplifies the calculation.

   Data: (1) The total mass coefficient for water ($\rho = 1.0$ g/cm^3) for a 1.25 MeV gamma-ray is 0.0632 cm^2/g, and similarly, for iron ($\rho = 7.874$ g/cm^3) for a 1.25 MeV gamma-ray is 0.0532 cm^2/g. (2) It is known that two gamma-rays are emitted each with almost 100% frequency, a 1.17 MeV and a 1.33 MeV for Co-60.

5. Consider the following radioactive decay chain where λ_i is the decay constant for nuclide i and R_i is the constant rate of production (e.g., nuclides per second) of nuclide i. Nuclide C decays by two modes: fraction f of the decays produce nuclide D, with the remaining decays (1–f) producing nuclide F. Nuclides E and F are stable.

   a. Write the set of simultaneous differential equations that describes the number densities of the nuclides as a function of time. (e.g., the differential equation for the rate of change of the concentration N of a nuclide that is not being replenished is $dN/dt = -\lambda N$.)

   b. Assuming at time zero, the concentration of nuclide A is N_{A0}, solve this differential equation for this nuclide concentration as a function of time.

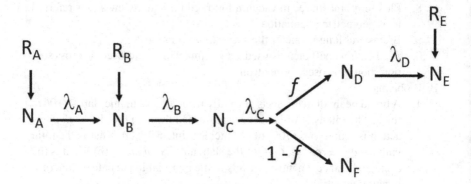

6. A mystery material has a density of 0.128×10^{24} at/cm^3, an absorption microscopic cross section $\sigma_a = 764$ b, and a total microscopic cross section $\sigma_t = 764$ b for neutrons with an energy of $E = 0.025$ eV.
   a. Calculate the *macroscopic* cross sections at 0.025 eV for absorption, scattering, and total interaction.
   b. What fractional attenuation will a 0.025-eV neutron *beam* experience when traveling through 1 mm of the material? 1 cm? That is, calculate $\phi(x)/\phi_0$. (c) Is the fractional attenuation calculated in part (b) based on the *actual* beam intensity? Explain why or why not?
7. A small, high-energy gamma-ray source is located in the vacuum of space. Ignoring background radiation, the gamma-ray flux at a distance of 1.5 m from the source is measured to be 2.40×10^{10} cm^{-2}-s^{-1}.
   a. What would be the gamma-ray flux due to the source at a distance of 3.0 m from the source?
   b. Repeat the problem but now assume that the measurements include the value of a constant *background* gamma-ray radiation, that is, a value of radiation that would be measured everywhere in the absence of the gamma-ray source. What is the background flux value in cm^{-2}-s^{-1} if the measured flux at 1.5 m is still 2.40×10^{10} cm^{-2}-s^{-1} and that at 3.0 m is 9.0×10^9 cm^{-2}-s^{-1}?
8. Compute the flux, macroscopic cross section, and reaction rate for the following data: $n = 2 \times 10^5$ cm^{-3}, $v = 3 \times 10^8$ cm/s, $N = 0.04 \times 10^{24}$ cm^{-3}, $\sigma = 0.5 \times 10^{-24}$ cm^2. Here, n is the particle number density, v the particle speed, and N the atom number density of the medium through which the particles are traveling.
9. At a particular location in a homogeneous material, the flux density of particles is 2×10^{12} cm^{-2}s^{-1}. (a) If the particles are photons, what is the number density of photons per cm^3 at that position? (b) If the particles are thermal neutrons (2200 m/s), what is the number density of neutrons per cm^3?
10. A broad beam of neutrons is normally incident on a homogeneous slab 6-cm thick. The intensity of neutrons transmitted through the slab without interaction is found to be 30% of the incident intensity. (a) What

is the total *macroscopic cross section* for the slab material *in cm⁻¹*?
(b) What is the average distance a neutron travels in the material in *cm*
before undergoing an interaction?

11. Calculate the half-thickness *in cm* for 1-MeV photons in (a) water,
(b) iron, and (c) lead. Here, *half-thickness* is the thickness of an infinite
slab of the material which results in half of incident beam photons penetrating the slab with no collisions.

Data at 1 MeV:

$(\mu/\rho)_{H2O} = 0.07066\,cm^2/g$, $(\mu/\rho)_{Fe} = 0.05951\,cm^2/g$, $(\mu/\rho)_{Pb} = 0.06803\,cm^2/g$.

9 Neutron Chain Reactions and Basic Nuclear Reactor Physics

OBJECTIVES

After studying this chapter, the reader should be able to:

1. Understand the basic of the chain reaction and distinguish differences and similarities between infinite and effective multiplication factors.
2. Identify thermal and fast reactor types and their basic functions.
3. Understand basic parts of nuclear reactors and their components toward reactor physics and mathematical formulas.
4. Understand the foundation of four-, five-, and six-factor formulas and use them to predict basic reactor power performance.
5. Use the criticality concept to estimate reactor size.

9.1 DEFINITIONS OF CHAIN REACTION AND NUCLEAR REACTOR[1]

The principle behind the neutron-induced fission reactor is the **chain reaction**. Regardless of the fissionable isotope, a neutron-induced fission produces on average two to three neutrons. If one of the emitted neutrons is captured by another fissionable nucleus, and that capture also results in fission, this fission in turn releases on average another two to three neutrons. A **chain reaction** results when the series of fission reactions becomes self-sustaining, that is, each fission triggers at least one more fission (see Figure 9.1). A **nuclear reactor** is a device in which a chain reaction is initiated, maintained, and *controlled*. Figure 9.2 shows the first man-made sustained chain reactions in Chicago 1942.

9.2 CRITICALITY AND MULTIPLICATION

The possibility of a chain reaction involving neutrons in a mass of nuclear fuel depends on: (1) the nuclear properties of the reactor such as cross sections and neutrons produced per absorption (i.e., *material properties*), and (2) the size, shape, and arrangement of the materials (i.e., *geometric properties*).

Criticality is attained when at least one of the several neutrons emitted in a fission process causes a second nucleus to fission. However, not all systems of fissionable material can go critical. In a mass of fissionable material, if more neutrons are lost by escape from the system (termed neutron **leakage**) or by non-fission absorption in

DOI: 10.1201/9781003272588-9

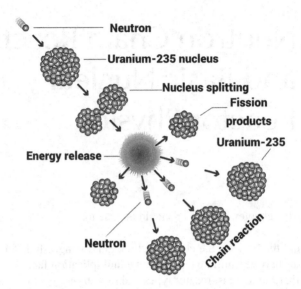

FIGURE 9.1 Schematic of a chain reaction.

FIGURE 9.2 The first man-made self-sustaining nuclear chain reaction initiated in Chicago Pile-1 (CP-1) on 2 December 1942.

impurities (termed **poisons**) than are produced in fission, then the chain reaction cannot be sustained. Such a system is said to be **subcritical**. It should be noted that more precisely, this term "leakage" should be labeled **net leakage** as some of the neutrons will undergo scattering collisions in the medium surrounding the reactor and as a result will be reflected back into the reactor where they are absorbed. A surrounding medium whose function is to enhance the scatter of leaked neutrons back into the rector is called a **reflector**.

If we have a sustained chain reaction, and if the rate of fission neutron production exceeds the rate of loss, the system is said to be **supercritical**. If we have a state of equilibrium in the neutron population, that is, exactly one neutron per each fission in turn initiates another fission, the system is said to be **critical**.

Another important term is **critical mass**, which is the minimum amount of mass of fuel of a system required to sustain a chain reaction, that is, go critical. Hereafter, the term **fuel** in general refers to all the uranium, plutonium, and thorium in the reactor.

A critical system is qualitatively described by the *neutron balance* relationship,

Rate of neutron production = Rate of neutron loss,

or, in separately accounting for each of the two neutron loss mechanisms,

Rate of neutron production = Rate of neutron absorption
+ Rate of neutron leakage.

9.3 MULTIPLICATION FACTORS

We can express the state of a chain reacting system (a reactor) by a single number k, termed the **multiplication factor**. The multiplication factor k for a chain reacting system, also termed the **effective multiplication factor**, k_{eff}, describes the neutron population growth of a system. In terms of the neutron balance equation,

k_{eff} = (rate of neutron production)/(rate of neutron absorption
+ rate of neutron leakage).

In general, $k_{eff} > 1$ indicates the system is *supercritical*, that is, the neutron population is increasing with time. If $k_{eff} = 1$, this implies that the system is *critical*, that is, the neutron population is constant with time. Lastly, $k_{eff} < 1$ indicates the system is *subcritical*, that is, the neutron population is decreasing with time. Since $k_{eff} = 1$ indicates criticality, the **excess neutron multiplication** is simply $(k_{eff} - 1)$. Therefore, $(k_{eff} - 1)/k_{eff}$ gives the *fractional excess neutron multiplication*; this term defines the system **reactivity**, ρ.

$$\rho = (k_{eff} - 1)/k_{eff} = \Delta k/k_{eff}$$

Here, ρ is a measure of the magnitude of a reactor's departure from criticality and ρ may be viewed as the fractional change in neutron population per neutron generation. The summary is given in Table 9.1.

The **infinite multiplication factor**, k_∞, is the value of the multiplication factor that would result if the system were infinite in size. That is, it's the multiplication factor ignoring neutron leakage from the system, and is a function of the material properties of the system alone. So, k_∞ is independent of geometry.

TABLE 9.1

Multiplication Factors and Reactivities as a Function of the State of Criticality

Effective Multiplication	Reactivity	Criticality State
$k_{eff} > 1$	$\rho > 0$	Supercritical
$k_{eff} = 1$	$\rho = 0$	Critical
$k_{eff} < 1$	$\rho < 0$	Subcritical

Example 9.1

An assembly containing fissionable materials is producing 5.2×10^5 n/s (neutrons per second), while 4.9×10^5 n/s are being absorbed and 6×10^4 n/s are leaking out of the reactor. Determine the infinite multiplication factor and calculate reactivity of this design.

Solution

If we ignore neutron leakage, then we can calculate the infinite multiplication factor,

$$k_\infty = \text{(neutron production)/(neutron absorption)}$$
$$= 5.2 \times 10^5 \text{ n/s}/4.9 \times 10^5 \text{ n/s} = 1.06.$$

This indicates that the material alone (assuming no neutron loss by leakage) is supercritical.

The effective multiplication factor denotes an overall subcritical condition.

$$k_{eff} = \text{(n production)/(n absorption + n leakage)}$$
$$= (5.2 \times 10^5 \text{ n/s})/(4.9 \times 10^5 \text{ n/s} + 6 \times 10^4 \text{ n/s}) = 0.95$$

The corresponding reactivity of the reactor is

$$\rho = (k_{eff} - 1)/k_{eff} = (0.95 - 1)/0.95 = \mathbf{-0.05}.$$

9.4 HOMOGENEOUS CONCEPT

Assume the reactor material properties can be characterized by single values of the macroscopic fission and macroscopic absorption cross sections; that is, the composition of the material is the same throughout the reactor. The material properties of such a system are everywhere **homogeneous**.

For the case of an **infinite, homogeneous** system, the neutron flux is the same everywhere, and the ratio of the fission reaction rate density to the absorption reaction rate density is given by $(\Sigma_f \phi)/(\Sigma_a \phi)$. As each fission reaction on average produces ν prompt fission neutrons, the ratio of the production rate of fission neutrons to the absorption rate (per unit volume) is thereby

$$(\nu \Sigma_f \phi)/(\Sigma_a \phi).$$

Since the flux is *constant* throughout an infinite, homogeneous system, the flux terms cancel, and we are left with the definition of the *infinite multiplication factor*:

$$k_\infty = \nu\Sigma_f/\Sigma_a.$$

Ignoring neutron leakage from a *finite*, homogeneous system, the reactivity of the core is then $\rho = (\nu\Sigma_f - \Sigma_a)/(\nu\Sigma_f)$. Note that as $k_{eff} \to \infty$, $\rho \to 1$. The equations for an infinite, homogeneous reactor, $k_\infty = \nu\Sigma_f/\Sigma_a$ and $\rho = (\nu\Sigma_f - \Sigma_a)/(\nu\Sigma_f)$ are extremely important and <u>must</u> be memorized.

Example 9.2

Assume the reactor fuel is metallic uranium (i.e., only consists of uranium metal as opposed to UO_2). The fast group constants for the fuel are given in the below table. If the fuel is 25 atom percent U-235, the rest being U-238, what is the fuel k_∞?

Material	ν	σ_f (b)	σ_γ (b)
^{235}U	2.6	1.4	0.25
^{238}U	2.6	0.095	0.16

Solution

Here, we have 25 a/o U-235, which means the U-238 atom density is related to that of U-235 by $N_{238} = 3N_{235}$. In general, the infinite multiplication factor for a mixture of present is

$$k_\infty = \frac{\sum_{i=1}^n \nu_i\Sigma_{fi}}{\sum_{i=1}^n \Sigma_{ai}},$$

where $i = 1$, and n indicates the different nuclides present. Note that the number of neutrons produced per fission ν is the same for both isotopes.

$$k_\infty = [\nu\Sigma_{f235} + \nu\Sigma_{f238}]/(\Sigma_{a235} + \Sigma_{a238})$$

$$= [\nu N_{235}\sigma_{f235} + \nu N_{238}\sigma_{f238}]/(N_{235}\sigma_{a235} + N_{238}\sigma_{a238})$$

$$k_\infty = [\nu N_{235}\sigma_{f235} + \nu(3N_{235})\sigma_{f238}]/(N_{235}\sigma_{a235} + 3N_{235}\sigma_{a238})$$

$$= [\nu\sigma_{f235} + 4\nu\sigma_{f238}]/(\sigma_{a235} + 4\sigma_{a238})$$

$$k_\infty = [(2.6)(1.4\,b) + 3(2.6)(0.095\,b)]/[(1.4\,b + 0.25\,b) + 3(0.095\,b + 0.16\,b)],$$

where in the denominator, $\sigma_a = \sigma_f + \sigma_\gamma$,

$$k_\infty = \frac{(3.64 + 0.74)}{(1.65 + 0.77)} = 1.81.$$

9.5 REACTIVITY CHANGE

By analogy to such concepts like enthalpy and potential difference, we are typically interested in the **change in reactivity** of a system as opposed to an absolute value of ρ.

That is, $\Delta\rho = (k_2 - k_1)/(k_1 k_2)$, where k_1 and k_2 are the effective multiplication factors corresponding to the initial and final states, respectively, of a reactor which has undergone a change in conditions. Reactivity is typically expressed as a unitless number. It is a property of the reactor core.

Example 9.3

What is the reactivity of a system with $k = 1.002$?

Solution

$$\rho = (k - 1)/k = (1.002 - 1)/1.002 = \mathbf{0.001996}$$

Example 9.4

k_{eff} for a reactor is calculated to be 1.025. After a change in fuel temperature, the reactor's k_{eff} is 1.027. What is the associated changed in reactivity?

Solution

$$\Delta\rho = (k_1 - k_2)/k_1 k_2 = (1.027 - 1.025)/(1.027 \times 1.025) = \mathbf{0.00190}$$

9.6 REACTOR NOMENCLATURE REVIEW

- **Nuclear reactor** can refer to the entire power plant or just to that portion where the nuclear fission is occurring—also known as the **reactor core** or simply the **core**. In this chapter, the term **reactor** will be assumed to be synonymous with **reactor core**.
- A **homogeneous** reactor is one where the reactor core's material composition is everywhere the same. So, the macroscopic cross sections have no location dependency.
- A **bare** reactor is one where any neutron that passes out of the core is assumed to not return. The boundary that is the surface of such a reactor is said to be *nonreentrant*.
- **Leakage** refers to the phenomenon of neutrons exiting the core.
- Most reactor designs have the core surrounded by a **reflector**. A reflector is typically composed of a material with a significant neutron scattering cross section such that many of the neutrons that leak from the core are scattered back into the core. Hence, a *bare* reactor is one that is not surrounded by a reflector.

Assuming a bare, homogeneous reactor that can be characterized by a single neutron energy group and the associated macroscopic cross sections,

$$k_{eff} = k_{\infty} P_{NL},$$

where P_{NL} is the **non-leakage probability** of a neutron from the core.

Example 9.5

Consider a *critical* reactor with a $k_{\infty} = 1.10$. What is the neutron non-leakage probability?

Solution

Here, we will use $k_{eff} = 1 = k_{\infty} P_{NL}$. Therefore, $P_{NL} = 1/k_{\infty} = 1/1.10 = \mathbf{0.909}$.

We can see that P_{NL} is dependent on the number of neutrons that leak from the core without returning. Leakage in turn depends on the frequency of scattering collisions and on the shape of the core, more specifically on the ratio of the core surface to volume because losses occur at the core boundary and production occurs within the core.

9.7 REACTOR CORE DESIGN AND THE IMPORTANCE OF k_{∞}

Students may wonder the point of calculating k_{∞} since it only applies to an infinite system—that is, one in which no neutron leakage occurs. The fuel in most existing reactors is in a solid form, the fuel being fabricated either as uranium dioxide fuel pellet, UO_2, or **mixed-oxide (MOX)** fuel pellets containing both uranium and plutonium fuel, UO_2-PuO_2. The fuel pellets are housed in cylindrical fuel elements (also called fuel rods or fuel pins) which are in turn grouped into fuel assemblies (which in some texts may also be referred to as fuel elements). The fuel assemblies are grouped together into a roughly cylindrical shape, thus creating the reactor core (see Figure 9.3).

Figure 9.4 shows the core loading plan for a PWR, the fuel assemblies are identified as members of specific fuel regions (also called fuel batches) based on the *enrichment* of the fuel at the time of fabrication. Most commercial power reactors replace about one-third of the assemblies in the core with freshly fabricated fuel assemblies during a refueling. Hence, most of the fuel assemblies in such a core can be classified as being *fresh* (regions 32A and 32B, denoted as "Feed" in the above diagram), *once-burned* (regions 31A and 32B), or *twice-burned* (region 30). Note that the fuel is loaded in a radially symmetric pattern.

The arrangement of the fuel in a reactor core is important in several ways: determining how long the core can be operated until refueling is required, ensuring that the operational characteristics of the core and its fuel meet required licensing limits, and providing good neutron economy, for instance, in limiting the net neutron leakage from the core.

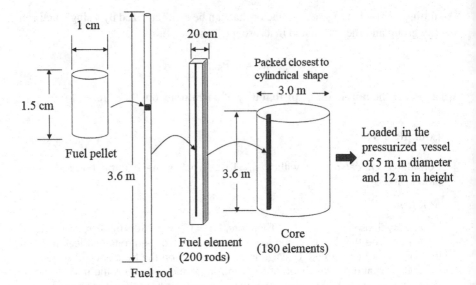

FIGURE 9.3   Construction of a reactor core.

The reactivity of any fuel assembly at the time of operation is dependent on its location in the core—that is, on the core geometry or loading pattern. In order to give core designers a better idea of the inherent reactivity of a fuel assembly independent of its eventual location in the core, the designers calculate the k_∞ value for the assembly.

9.8   REACTOR CLASSIFICATION

There are a number of different schemes for classifying reactors such as the dominant energy spectrum (fast or thermal), the type of moderator (light water, heavy water, or graphite), the temperature (high temperature), and so forth. Herein, we shall consider the classification of reactors by the dominant neutron energy spectrum as either **thermal** reactors or **fast** reactors (see Figure 9.5). Recall that for *fissile* isotopes, the microscopic absorption cross section increases substantially for neutrons in the *thermal* energy range—that is, neutrons with kinetic energies on the order of 1 eV or less. In the thermal energy range, the microscopic absorption cross section for a fissile isotope is approximately proportional to 1/v, that is, one over the neutron speed (refer back to Figure 8.3).

Thermal reactors contain a substantial amount of low-atomic-weight material(s) in order to effectively cause the high-energy prompt fission neutrons to rapidly lose energy from elastic scatter collisions and to eventually slow down to thermal energies, that is, to energies below about 1 eV. Such materials are called neutron **moderators**. The most common moderating materials used in nuclear reactors are ordinary water, heavy water, and graphite. In light water reactors (LWRs) and heavy water reactors (HWRs), the water fulfills dual functions, both as the neutron moderator and the coolant that removes the heat from the reactor core. Thermal reactors can also be designed with a gas coolant

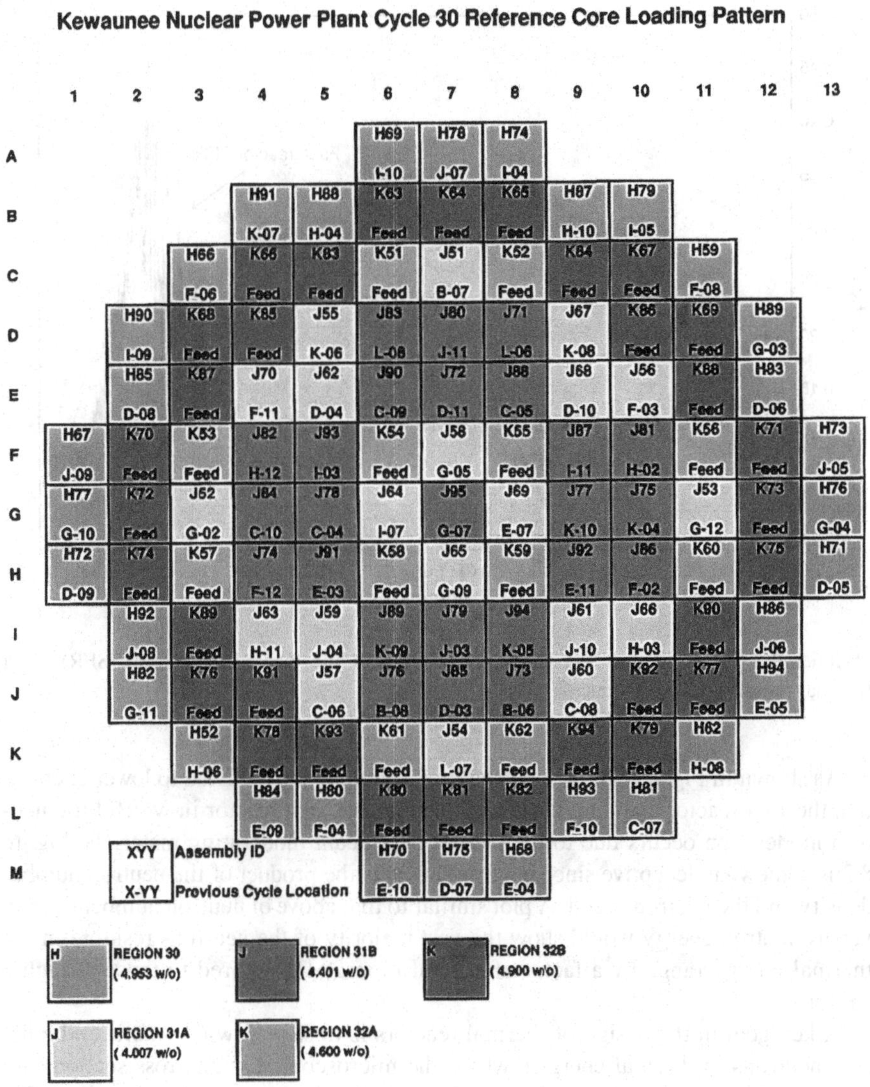

Kewaunee Nuclear Power Plant Cycle 30 Reference Core Loading Pattern

FIGURE 9.4 Core loading pattern of a commercial pressurized water reactor (PWR). This represents a radial cross section of the reactor core with each colored square a single fuel assembly.

such as helium, in which case the moderator is typically graphite. The process of slowing high-energy prompt fission neutrons to thermal energies is called **thermalization**. Above the thermal energy range, the target nuclei can typically be assumed to be at rest with respect to the neutrons. At thermal neutron energies, the motion of the target nuclei relative to the colliding neutron must be taken into account—in other words, the neutrons are in a state of approximate thermal equilibrium; therefore, their distribution in energy space roughly follows the characteristic of a Maxwellian distribution.

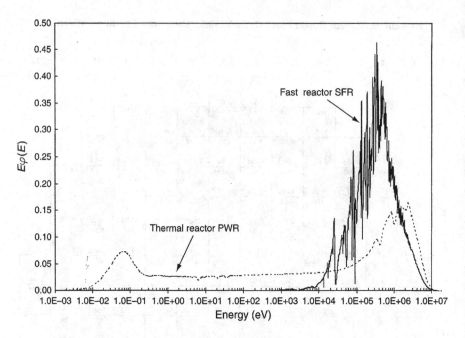

FIGURE 9.5 Neutron flux energy distribution in a sodium-cooled fast reactor (SFR) and a thermal pressurized water reactor (PWR).

As shown in Figure 8.3, the neutron flux is significantly shifted to lower energies in a thermal reactor compared to a *fast* reactor; that is, a reactor in which little neutron moderation occurs due to the lack of significant moderating materials. Figure 8.3 is somewhat deceptive since neutron flux ϕ is the product of the neutron number density and the neutron speed. A plot similar to that above of neutron number density versus neutron energy would show the vast majority of the neutrons residing in the thermal energy range by a factor of several thousand compared to those at higher energies.

A key goal in the design of thermal reactors is to slow down (*moderate*) the fission neutrons to thermal energies where the microscopic fission cross sections are significantly greater. In the thermal energy region, only the *fissile isotopes* ^{233}U, ^{235}U, ^{239}Pu, and ^{241}Pu are useful for causing fissions. Values given for thermal neutron microscopic cross sections typically correspond to a neutron energy of 0.0253 eV (2200 m/s).

9.9   FAST AND THERMAL FLUX

In two-group neutron transport theory, the neutron spectrum is divided into two energy groups, those of fast neutrons with kinetic energies above 1 eV, and those of thermal neutrons with kinetic energies of 1 eV or less. Fast neutron parameters are typically represented by subscript 1 and thermal neutron parameters by subscript 2. Assuming that the core average fast and thermal neutron fluxes are equal, we have,

$$\phi_1 = n_1 v_1 = \phi_2 = n_2 v_2,$$

where n represents the neutron number density and v the average neutron speed. Rearranging, we have

$$\phi_2/\phi_1 = 1 = (n_2/n_1)(v_2/v_1) \rightarrow n_2/n_1 = v_1/v_2.$$

The neutron speed is proportional to the square root of the neutron energy since for reactor neutrons, the kinetic energy $E = \frac{1}{2} mv^2$.

$$n_2/n_1 = [E_1/E_2]^{1/2}$$

Substituting typical fast and thermal neutron energies of 1 MeV and 0.025 eV, we have

$$n_2/n_1 = [10^6 \text{ eV}/0.025 \text{ eV}]^{1/2} = 6320,$$

so, in a thermal reactor, the thermal neutron population outnumbers the fast neutron population by a factor of several thousand.

9.10 FAST REACTORS

Neutron distribution in high energy range is shown in Figure 9.6. In fast reactor design, the goal is to have neutrons with kinetic energies significantly above thermal cause fission. As seen in Figure 7.7 in Chapter 7, the majority of prompt fission neutrons are born with kinetic energies above several keV. The fast fission neutrons undergo minimum slowing down before they are absorbed by the fuel and cause the production of a new generation of fast fission neutrons. Although the absorption cross sections (including fission) for high-energy fission neutrons are several hundred times smaller than those for thermal neutrons with fissile isotopes, at fast energies, practically all reactor fuel fissions (i.e., U, Pu, and Th isotopes). To prevent any significant neutron moderation from occurring, fast reactor designs limit the use of materials with low atomic weights. This rules out the use of water, either ordinary or heavy, from being used to remove reactor heat. Most fast reactor

FIGURE 9.6 Oklo site.

coolants are either liquid metals, the two most popular being sodium and lead, or a gas such as helium.

9.11   ENRICHMENT AND OKLO

In general, **enrichment** refers to the relative abundance of an isotope in some material and is frequently used to indicate an abundance of an isotope greater than that found in nature. Because a reactor chain reaction cannot be maintained by a fuel composed only of natural uranium with ordinary water as a moderator, the enrichment of the fissile U-235 in such reactors (referred to as LWRs or light water reactors) must be increased above that found in nature. Enriched uranium therefore refers to uranium in which the abundance of U-235 has been increased above the natural 0.711 weight percent. Most fission reactors, commercial, naval, and research, in the world today are LWRs. The commercial LWRs typically use uranium fuel enriched to about three to five weight percent.

More generally, with regard to nuclear fuel, we can view **enrichment** as the weight (mass) percent of the fuel that is *fissile*. Here, we define **fuel** as being the *fissionable* material in the reactor. This is essentially all the uranium, plutonium, and thorium in the reactor core.

As you may recall from the earlier homework problem comparing the radioactive decay of U-235 to U-238, because of their different half-lives, the natural abundance of U-235 in uranium was about 3 weight percent roughly 1.8 billion years ago. This was a high enough enrichment to maintain a sustained chain reaction in uranium in the presence of a moderator such as ordinary water. Evidence for this actually exists at a place called Oklo in Gabon, Africa (see Figure 9.6). More than 1.5 billion years ago, fission reactions repeatedly occurred in an underground uranium deposit over hundreds of thousands of years. Apparently contact with ground water resulted in the moderation of neutrons from spontaneous fission to the extent that criticality occurred. The evidence was discovered in 1972 when fission products were found in the deposit. The fact that the fission products are still in place in the rock matrix after over 1.5 billion years provides information in support of the long-term viability of the burial of high-level radioactive waste.

9.12   NEUTRON LEAKAGE, SIZE, AND CRITICALITY

The relationship between the actual or effective multiplication factor (k_{eff}) of a reactor core and its infinite multiplication factor (k_∞) is given by $k_{eff} = k_\infty P_{NL}$, where P_{NL} is the neutron non-leakage probability for the core. In the case of a *bare* reactor core, that is, one that be assumed to be surrounded by a vacuum, any neutrons exiting the core do not return since no material attenuation, in particular, neutron scattering, occurs outside of the reactor core. A comparison of two identical reactor cores, one bare and the other with a reflector, will always result in the reflected core having a higher reactivity (see Figure 9.7).

Consider a bare reactor core that was critical. If a person approached the core and stood next to it, the core condition would change to supercritical since some of the escaping neutrons would interact with the nuclei of the atoms inside the person's body and be scattered back into the core.

"Bare" Reactor Core   "Reflected" Reactor Core

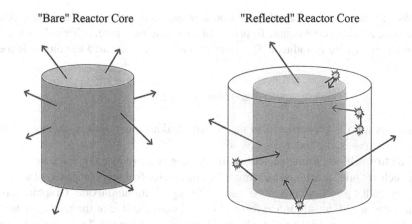

FIGURE 9.7   Comparison of two identical reactor cores. The reactivity of the core with the reflector (right) will always be higher than that of the bare core since the reflector will scatter some of the neutrons back into the core, thereby increasing its neutron population over that of the bare core.

Neutron leakage is proportional to the surface area to volume ratio of the reactor.

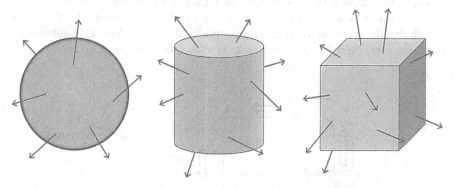

For equal volume shapes, the sphere has the lowest surface area to volume ratio.

FIGURE 9.8   The reactivity of reactors of identical volume and composition will vary according to the size of their respective surface areas.

Since P_{NL} is a probability, its value must be less than or equal to one. For an infinite reactor core, $P_{NL} = 1$, and $k_{eff} = k_\infty$, otherwise, $k_{eff} < k_\infty$. Consider reactors of identical k_∞ volume. The difference in reactivity values between such reactors will depend entirely on their relative surface areas, that is,

Leakage $\propto$ (reactor surface area)/(reactor volume).

Since the sphere has the lowest ratio of surface area to volume of any geometric shape, it in turn will have the lowest leakage, all else being equal. Hence, if the shape of a *critical* spherical reactor core could be deformed, its reactivity condition would immediately change from critical to subcritical since any geometric shape has a higher surface area-to-volume ratio than the sphere (see Figure 9.8).

We can now demonstrate that neutron leakage decreases with increasing reactor core size, all else being equal. To prove this, we assume a bare, spherical, homogeneous, reactor core of radius r. The ratio of the core surface area to volume is then given by

$$(4\pi r^2)/(4\pi r^3/3) = 3/r.$$

Since this ratio is proportional to the neutron leakage, as r, that is, the size of the reactor increases, the leakage decreases.

The first self-sustaining nuclear chain reaction was produced by the construction of a nuclear "pile" reactor (see Figure 9.2). The reactor fuel was composed of approximately 5640 kg of uranium metal and 36,300 kg of uranium oxide. Graphite moderator was added, mostly in the form of bricks, until the size of the reactor reduced the neutron leakage to the point where criticality was achieved. This required about 350,000 kg of graphite.

The operation of a fluid-fuel reactor is based on the same concept. Unlike most reactors where the fuel is in a solid form, typically a uranium dioxide ceramic, UO_2, or an MOX fuel composed of both UO_2 and PuO_2, in a fluid-fueled system the fuel is part of the coolant. The most famous of such designs is that of the Molten Salt Reactor Experiment at Oak Ridge National Laboratory from 1965 to 1969.

As shown in Figure 9.9, the size and geometry of the system outside of the core is such that the excessive neutron leakage results in the fluid-fuel being subcritical. Inside the core, the reduced neutron leakage allows the fluid-fuel to go critical.

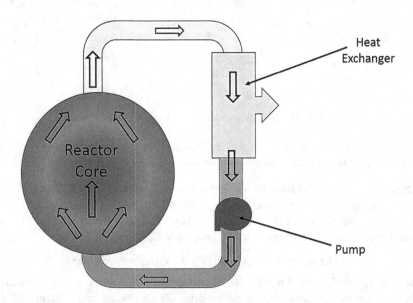

FIGURE 9.9 The molten salt reactor principle where the decreased neutron leakage of the fluid-fuel inside the core results in momentary criticality. Conversely, the fluid-fuel is subcritical when in the higher leakage piping.

9.13 RECALL ON THERMAL NEUTRON

In other to start the concept on fundamental reactor theory, we have to go back to the thermal neutrons, which can be described by a Maxwellian velocity distribution:

$$n(v) = n_0 \frac{4\pi v^2}{\left(\frac{2\pi kT}{m}\right)^{3/2}} \exp\left(-\frac{mv^2}{2kT}\right),$$

where $n(v)$ is the number of neutron per cm^3 per unit velocity v (m/s) interval, n_0 is the thermal neutrons per cm^3, m is the neutron rest mass (kg), T is the temperature (K), and k is the Boltzmann constant ($1.38 \times 10^{-23} m^2 kg\, s^{-2}\, K^{-1}$). Maxwellian velocity distribution of neutron under various temperature conditions is shown in Figure 9.10.

By taking a derivative of the above equation with respect to velocity and set it to zero, it is possible to find the **most probable neutron velocity**, v_p, which is

$$v_p = \left(\frac{2kT}{m}\right)^{1/2},$$

and the kinetic energy of neutrons at the most probable velocity (KE_p) is given by

$$KE_p = \frac{mv_p^2}{2} = \frac{m}{2}\left(\frac{2kT}{m}\right) = kT.$$

This is not the average kinetic energy, which has the value of $\frac{3}{2}kT$ and it is independent of the particle mass. For neutrons at 293 K (20°C), we can see that

$$v_p = \left(\frac{2 \times 1.38 \times 10^{-23} \times 293}{1.66 \times 10^{-27}}\right)^{1/2} = 2200 \text{ m/s}.$$

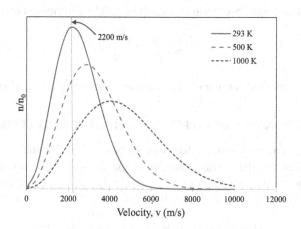

FIGURE 9.10 Maxwellian velocity distribution of neutron at different temperatures.

and this is shown in Figure 9.10. At this condition, the most probable kinetic energy is calculated to be roughly 0.025 eV.

$$KE_p = kT = \left(1.38 \times 10^{-23}\, \frac{J}{K}\right)(293\ K)\left(\frac{1\ eV}{1.6 \times 10^{-19} J}\right) = 0.02527\ eV.$$

Applying an integration technique, it can be shown that the average neutron velocity can be expressed as

$$\langle v \rangle = \frac{\displaystyle\int_0^\infty n(v)v\,dv}{\displaystyle\int_0^\infty n(v)\,dv} = \sqrt{\frac{8kT}{\pi m}}.$$

Thus, the ratio of $\langle v \rangle / v_p$ is $2/\sqrt{\pi} = 1.128$.

9.14 CORRECTED ABSORPTION CROSS SECTIONS

In the thermal region, the absorption cross section will be proportional to the inverse of velocity ($\sigma_a \propto 1/v$). Using the idea from the previous frame, we can see that it will vary inversely as the square root of both the kinetic energy and temperature.

$$\sigma_a \propto \frac{1}{v} \propto \frac{1}{KE^{1/2}} \propto \frac{1}{T^{1/2}}$$

We can see that the absorption cross section at a temperature T will be

$$\sigma_a(T) = \sigma_{a\,293}\sqrt{\frac{293}{T}}.$$

This is referenced to the absorption cross section at 293 K. Since the average neutron velocity is greater than the most probable neutron velocity by a factor of $2/\sqrt{\pi}$, it means that the average absorption cross section can be determined by using

$$\sigma_a(T) = \sigma_{a\,293}\frac{\sqrt{\pi}}{2}\sqrt{\frac{293}{T}}.$$

This correction method is often used in nuclear engineering calculation.

9.15 RELATIONSHIP BETWEEN MODERATOR AND LETHARGY

We have learned already about lethargy in Section 7.15 in Chapter 7. In that lesson, we can calculate N, the average number of elastic scatters required to reduce a neutron of one energy to a lower energy. So, if we let E_n to be the average thermal neutron energy, then we can determine the number of collisions to thermalize; that is,

$$\text{\# of collisions to thermalize} = \ln(E_0/E_n)/\xi.$$

The log energy decrement does not fully describe the excellence of a material as a moderator. Because we would want a high probability for scatter and a low probability of neutron being absorbed. So, we have to find more ways into determining a good moderator.

9.16 MACROSCOPIC SLOWING-DOWN POWER

Here, we define the **macroscopic slowing-down power** (MSDP) as the product of lethargy and the macroscopic scattering cross section for epithermal neutrons.

$$\text{MSDP} = \xi \Sigma_s^{\text{epi}}.$$

This will tell you how rapidly slowing down will occur in the material. We don't want to have a low MSDP value. For a light gas like helium-4, lethargy will be good but MSDP will be poor because of the small probability of scatter in the epithermal region. Hence, the microscopic scatter cross section is small and the atomic density of helium is also small. For a compound or a mixture (assuming three components), it is important to point out that lethargy can be found by taking an average as follows:

$$\xi = \frac{\xi_1 \Sigma_{s1}^{\text{epi}} + \xi_2 \Sigma_{s2}^{\text{epi}} + \xi_3 \Sigma_{s3}^{\text{epi}}}{\xi_1 + \xi_2 + \xi_3}.$$

And the last parameter to help us determine the moderator is …

9.17 MODERATING RATIO

Moderating ratio (MR) is required because MSDP alone cannot describe or provide a complete picture of an effective moderator. For example, boron has a high lethargy value and a good MSDP. However, it is still a poor moderator because of its high probability of absorbing neutrons. So, we define the **moderating ratio (MR)** as a ratio of MSDP to the macroscopic absorption cross section in the thermal region.

$$\text{MR} = \frac{\xi \Sigma_s^{\text{epi}}}{\Sigma_a^{\text{th}}}$$

9.18 HOW TO DETERMINE A GOOD MODERATOR?

A good moderator must have low lethargy, high MR, and high MSDP. Table 9.2 shows various characteristics of moderator candidate materials.

Results shown in Table 9.2 indicate that water is an adequate candidate based on the values of lethargy, MSDP, and MR. The other aspects to concern are the material's availability and cost; in this case, water is an appropriate moderator. We can see that heavy water has an excellent MR value but it is expensive. Beryllium and carbon (i.e., graphite) are comparable in MR value. But beryllium (Be) is difficult to fabricate, slightly expensive, and has high toxicity. Helium (He) is often ruled out

TABLE 9.2

Comparison of the Moderating Characteristics of Materials

Materials	Lethargy, ξ	# Collision to Thermalize	MSDP	MR
Water, H_2O	0.927	19	1.425	62
Heavy water, D_2O	0.510	35	0.177	4830
Helium, He	0.427	42	8.87×10^{-6a}	51
Beryllium, Be	0.207	86	0.1538	126
Boron, B	0.171	105	0.092	0.00086
Carbon, C	0.158	114	0.083	216

[a] At 1 atm and 293 K.

due to its density and extreme low MSDP value. Boron, on another hand, is a poor moderator due to its high macroscopic absorption cross-section value; so, it is often used as a neutron absorber in control rods.

9.19 AVERAGE VALUE OF THE COSINE OF THE SCATTERING ANGLE

From the discussion of neutron scattering in Section 7.12 and 7.13 in Chapter 7, the derivation and analysis were based on the **center-of-mass system**, where we would pretend to be the observer that was moving at the speed and in the direction that the compound nucleus would have after the collision. Also, in the elastic scattering, we can take another perspective where both observer and the target nucleus are both stationary; this is known as the **laboratory system**. Figure 9.11 shows the sketch of both systems.

Here, the important variables in our discussion are ψ and θ, which are the scattering angle in lab system of emergent neutron with respect to original direction of motion and neutron scattering angle in center-of-mass system, respectively. When we combine the velocity of compound nucleus with the velocity of scattered monoenergetic neutrons in the center-of-mass system, the scatter in the laboratory system becomes anisotropic, as shown in Figure 9.12. This shows a preferential forward scatter.

Thus, the average value of cosine of θ can be found and shown to be zero. It implies that θ is 90° and the scatter is isotropic in nature. We can use the Pythagorean theorem to apply to Figure 9.11 to find the relationship between ψ and θ, which is displayed in Figure 9.13. Based on this information, it can be shown that

$$\cos(\Psi) = \frac{A\cos(\theta) + 1}{\sqrt{A^2 + 2A\cos(\theta) + 1}}.$$

From the above equation, after a few mathematical steps, we can derive the average value of the scattering angle in the laboratory system, which is

$$<\cos(\Psi)> = \frac{2}{3A}.$$

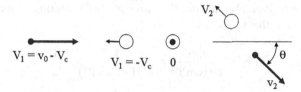

Laboratory system

Center-of-mass system

FIGURE 9.11 Scattering process in both laboratory and center-of-mass systems.

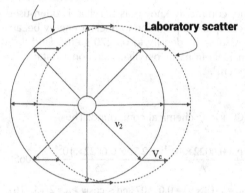

FIGURE 9.12 Illustration an anisotropic scatter in laboratory system from isotropic center-of-mass scatter.

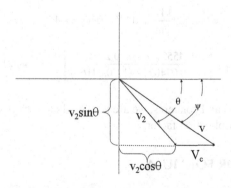

FIGURE 9.13 Relation between center-of-mass and laboratory systems.

For graphite, $<\cos(\psi)> = 2/(3 \times 12) = 0.056$ and $\psi \approx 87°$, which is nearly an isotropic scatter. However, for ^1H, the average value is 2/3 and $\psi \approx 48°$ showing a strong forward scatter.

9.20 TRANSPORT MEAN FREE PATH

The **transport mean free path** (mfp_{tr}) represents a scattering mean free path after being corrected for a traveling distance in the laboratory system due to the forward scatter preference. It can be calculated by dividing the scattering mean free path by the adjusted factor; that is,

$$mfp_{tr} = \frac{mfp_s}{1 - \langle \cos\Psi \rangle} = \frac{1}{\Sigma_s \left(1 - \langle \cos\Psi \rangle\right)} = \frac{1}{\Sigma_{tr}}.$$

We can see that its reciprocal is indeed the macroscopic transport cross section.

Example 9.6

Beryllium oxide, commonly known as Beryllia, is often used in a wide range of applications from space exploration to nuclear energy because of its ideal refractory material. Assume its density to be 270 g/cm³. Find the transport mean free path for thermal neutrons in beryllia. Additional data: $\sigma_s^{Be} = 7 \times 10^{-24} cm^2$ and $\sigma_s^O = 4.2 \times 10^{-24} cm^2$.

Solution

Beryllia is "BeO" for its chemical compound. Thus,

$$N_{Be} = N_O = \frac{\rho \times 0.6022 \times 10^{24}}{A} = \frac{2.70 \times 0.6022 \times 10^{24}}{25} = 0.06503 \times 10^{24} \text{ atoms/cm}^2$$

So, $< \cos\psi >_{Be} = 2/(3 \times 9) = 0.07407$ and $< \cos\psi >_O = 2/(3 \times 16) = 0.04167$.

We can calculate the macroscopic scattering cross sections for Be and O.

$$\Sigma_s^{Be} = 0.6503 \times 7 = 0.45521 \frac{1}{cm} \text{ and } \Sigma_s^O = 0.6503 \times 4.2 = 0.27313 \frac{1}{cm}$$

$$\text{Therefore, } mfp_{tr} = \left[\frac{0.45521}{(1 - 0.07407)} + \frac{0.27313}{(1 - 0.04167)} \right]^{-1} = (0.6832)^{-1} = 1.46 \text{ cm}.$$

With these definitions and concepts, we are now ready to introduce a detailed definition of effective multiplication factor.

9.21 SIX-FACTOR FORMULA

$$k_{eff} = \varepsilon p f \eta P_{NL}^{th} P_{NL}^{f}$$

where ε is the fast fission factor, p is the resonance escape probability, f is the thermal utilization factor, η is the thermal fission factor (also known as reproduction factor), and P_{NL} is the non-leakage probability for thermal (superscript th) and fast (superscript f) region. If one combines the non-leakage probability terms together into one term, the above equation becomes the **five-factor formula**. And if we ignore the non-leakage terms, that is, an infinite medium (i.e., independent of geometry), then the equation becomes the **four-factor formula**, and can be expressed as

$$k_\infty = \Sigma p f \eta.$$

The four-, five-, and six-factor formulas have been one of the earliest methods to model thermal reactor. It helps breaking neutron behavior into four to six parts. We now will discuss each variable in more detail.

9.22 FAST FISSION FACTOR (ε)

$$\varepsilon = \frac{(\text{\# of fast neutron from thermal} + \text{fast fissions})}{(\text{\# of fast neutrons from thermal fissions ONLY})}.$$

Its purpose is to account for fast fissions, mainly from uranium-238 and thorium-232. It is quite difficult to calculate accurately because these fertile isotopes may prevent neutrons from reaching thermal energies via transmutation and both fertiles have fission cross section at energy higher than 1 MeV. Typically, the value of ε ranges between 1.02 and 1.08 for natural low enriched uranium fuel, and it is typically 1 for highly enriched reactor fuels. In that sense, the total number of fast neutron is $n\varepsilon$; that is,

$$\text{Total fast neutrons} = n\varepsilon.$$

9.23 RESONANCE ESCAPE PROBABILITY (P)

$$p = \frac{\text{\# of neutrons thermalized}}{\text{total \# fast neutrons}}$$

This is the probability that a neutron will escape resonant capture and will reach thermal energies. We can relate this to

$$\text{Number of thermal neutrons} = n\varepsilon p.$$

It is a bit tricky to calculate p. In general, for the fuel with U-238 being present and that σ_a^{238} and N^{238} represent information of resonance absorbers,

$$\left(\sigma_a^{238}\right)_{eff} = \frac{\sigma_a^{238}}{\dfrac{N^{238}\sigma_a^{238}}{\Sigma_a}+1},$$

and p can be expressed as

$$p = \exp\left[-\frac{N^{238}}{\xi\Sigma_s} \int_{E_{th}}^{E_0} \left(\sigma_a^{238}\right)_{eff} \frac{dE}{E} \right],$$

where $\int_{E_{th}}^{E_0} \left(\sigma_a^{238}\right)_{eff} \frac{dE}{E}$ is the effective resonance integral (I_{eff}) and can be approxi-

mated by

$$I_{eff} = 3.9 \left(\frac{\Sigma_s}{N^{238}} \right)^{0.415},$$

where the unit of I_{eff} is expressed in barn. This equation satisfies for ratios of Σ_s/N^{238} less than 1000 barns. The range is 9.25 b for pure metal of U-238 to 240 b for an infinitely dilute mixture of U-238 in moderator. Then, p can be approximated by

$$p = \exp\left(-\frac{N^{238}}{\xi\Sigma_s} I_{eff} \right).$$

9.24   THERMAL UTILIZATION FACTOR (F)

$$f = \frac{\text{\# of neutrons absorbed in fuel}}{\text{total number of neutrons aborbed in fuel, mderator, cladding, etc.}}$$

Based on this definition, $f = 0$ if there is no fuel in the core and $f = 1$ for 100% fuel in the core. So, we can further express it mathematically using the macroscopic absorption cross sections for fuel and non-fuel type; that is,

$$f = \frac{\left(\Sigma_a\right)_{fuel}}{\left(\Sigma_a\right)_{fuel} + \left(\Sigma_a\right)_{non-fuel}} = \frac{\sigma_{a,\,fuel}}{\sigma_{a,\,fuel} + \sigma_{a,\,non-fuel}\left(\dfrac{N_{non-fuel}}{N_{fuel}}\right)}.$$

So, the thermal neutrons absorbed in fuel $= n\varepsilon pf$.

9.25   THERMAL FISSION FACTOR (η)

$$\eta = \frac{\text{\# of fast neutrons from thermal fission}}{\text{\# of thermal neutrons absorbed in fuel}}$$

Based on this physical description, it can be mathematically as

$$\eta = \nu \frac{\sigma_f}{\sigma_a} = \nu \frac{\Sigma_f}{\Sigma_a},$$

TABLE 9.3

Estimated Values of ν and η for Common Fissile Elements in Thermal Energy

Fissile	ν	η	Comments
U-233	2.49	2.29	Highest η for fuel, daughter of ^{232}Th
U-235	2.43	2.07	Lowest η for fuel, but naturally existed
Pu-239	2.87	2.11	Good η and high ν for fuel, daughter of ^{238}U
Pu-241	2.92	2.15	Highest ν for the fuel, but not naturally existed

where ν is the number of fast neutrons emitted per fission. Table 9.3 provides general values for ν and η for various fissile materials for nuclear reactors.

Therefore, the total number of fast neutrons in the next generation (n*) can be determined by using the following relationship, which is

$$n^* = n\epsilon pf\eta = nk_\infty.$$

9.26   PHYSICAL MEANING OF FOUR-FACTOR FORMULA

What we see from the previous frame is the four-factor formula,

$$k_\infty = \frac{n^*}{n} = \epsilon pf\eta.$$

And this can be described by the physical meaning of neutron (n) for each parameter mathematically; that is, in respective order of parameters,

$$k_\infty = \left(\frac{\text{fast n}}{\text{fast n from thermal fission}}\right) \times \left(\frac{\text{thermal n}}{\text{fast n}}\right) \times \left(\frac{\text{thermal n absorbed in fuel}}{\text{thermal n}}\right)$$
$$\times \left(\frac{\text{fast n from thermal fission}}{\text{thermal n abosrbed in fuel}}\right).$$

9.27   THERMAL NON-LEAKAGE PROBABILITY ($P_{NL}{}^{TH}$)

$$P_{NL}^{th} = \frac{1}{1+L^2B^2},$$

where L^2 is the **thermal diffusion area** (typically in cm^2) and B^2 is the **critical buckling** of a reactor. We will stay very basic here and leave the derivation of B^2 and other important aspects for the nuclear reactor theory course. In general, B^2 depends on reactor geometry and core size. Several important expressions for B^2 for simple

TABLE 9.4

Critical Flux Profiles and Bucklings for Several Homogeneous Bare Assemblies. Origin Is Taken at the Assembly's Center

Geometry	Dimensions	Flux profile	B^2
Slab	Thickness c	$A\cos\left(\dfrac{\pi x}{c}\right)$	$\left(\dfrac{\pi}{c}\right)^2$
Sphere	Radius R	$\dfrac{A}{r}\sin\left(\dfrac{\pi r}{R}\right)$	$\left(\dfrac{\pi}{R}\right)^2$
Cylinder	Radius R, Height H	$AJ_0\left(\dfrac{2.405r}{R}\right)\cos\left(\dfrac{\pi z}{H}\right)$	$\left(\dfrac{2.405r}{R}\right)^2 + \left(\dfrac{\pi}{H}\right)^2$

Note: J_0 is the Bessel function.

core geometries are given in Table 9.4. One important observation is that as the core increases in size, $B^2 \to 0$ and $P_{NL}^{th} \to 1$, implying that there is no leakage.

Here, the thermal diffusion area can be calculated by

$$L^2 \equiv \frac{D}{\Sigma_a},$$

where D is the **thermal diffusion coefficient** (in cm) and Σ_a is the thermal macroscopic absorption cross section of the reactor core material. D is directly proportional to the transport mean free path, which is

$$D = \frac{1}{3\Sigma_{tr}} = \frac{1}{3\Sigma_s(1- <\cos\Psi>)} = \frac{mfp_{tr}}{3}.$$

Table 9.5 provides a list of D, Σ_a, and L^2 for several pure moderators.

9.28 FAST NON-LEAKAGE PROBABILITY (P_{NL}^F)

This is related to the leakage of fast neutrons from the core. It is the probability a fast neutron does not leak from the core as it slows down to thermal energies. It can be estimated by using the following expression:

$$P_{NL}^f = \exp\left(-B^2\tau\right),$$

where τ is the **Fermi age** to thermal energies, and it is $\approx 1/6$ the mean squared distance between production and thermal energies. τ values are given in Table 9.5 for several moderating materials. Once again, we can see that as the core size increases, $B^2 \to 0$, and the non-leakage probability approaches to 1.

TABLE 9.5

Properties of Several Moderators for Thermal Neutrons (0.00253 eV)

Material	Density (g/cm³)	Σ_a (1/cm)	D (cm)	L (cm)	τ (cm²)
H_2O	1.00	0.0197	0.16	2.85	27
D_2O	1.10	0.000029	0.87	170	131
Be	1.85	0.00104	0.50	21	102
BeO	2.96	0.00060	0.47	28	100
C	1.60	0.00024	0.84	59	368

ANL-7010 report.

9.29   THE EXISTENCE OF FIVE-FACTOR FORMULA

We already know that when we combine the non-leakage probability parameters for both thermal and fast neutrons into one parameter, then the six-factor formula becomes the five-factor formula. We will describe this event and scenario by using a general analogy and we will leave the complicate derivations to the reader to learn more in the nuclear reactor theory course.

Here, under the two-group theory, we can approximate the fast and thermal flux. After a long derivation, it is possible to show that the effective multiplication factor can be expressed as

$$k_{eff} = \frac{\varepsilon p f \eta}{\left(B^2 L_f^2 + 1\right)\left(B^2 L_{th}^2 + 1\right)} = \frac{1}{\left(B^2 L_f^2 + 1\right)} \frac{1}{\left(B^2 L_{th}^2 + 1\right)} k_\infty,$$

where subscript f stands for fast neutrons and th stands for thermal neutron. Thus, we can see that the non-leakage probability can be roughly estimated by

$$P_{NL} = \frac{1}{\left(B^2 L_f^2 + 1\right)} \frac{1}{\left(B^2 L_{th}^2 + 1\right)}.$$

Definitely, the last term on the RHS is identical to the $P_{NL}{}^{th}$. And L_f is often described by the Fermi age. For large reactors, we can simply replace the diffusion length L by the **migration length** M without substantial loss of accuracy. So, the above equation yields

$$P_{NL} = \frac{1}{1 + B^2\left(L_f^2 + L_{th}^2\right) + L_f^2 L_{th}^2 B^4}.$$

We can observe that the value of buckling will be extremely small for large reactor core. Thus, B^4 term can be dropped out. Thus, the total non-leakage probability yields

$$P_{NL} = \frac{1}{1 + B^2 M^2},$$

where M^2 is the **migration area**, which is defined as $M^2 = L_f^2 + L_{th}^2$. By using M instead of L, this is often referred as *modified one-group theory*. Thus, you will end up with just the five-factor formula.

9.30 LIFE CYCLE OF NEUTRONS

We can ignore the leakage terms for now and inspect a life cycle of neutrons in an infinite reactor (i.e., independent of geometry). We can start with an initial number of neutrons (n) of 1000 with $\varepsilon = 1.04$, $p = 0.667$, $f = 0.90$, and $\eta = 2.22$. First, we can calculate the value of k_∞ to be $(1.04)(0.667)(0.90)(2.22) = \mathbf{1.3859}$. This is supercritical and in fact at the end of the cycle, there will be 1385.9 neutrons for the next cycle, that is, $n^* = 1385.9$. This can be illustrated by using the following logical thinking concepts (as shown in Figure 9.14).

Figure 9.15 shows a similar neutron life cycle to Figure 9.14, but with neutron diffusion added for both fast neutrons, as they slow down through the intermediate/ epithermal regions, as well as for thermal neutrons. This is analogous to the use of non-leakage probability.

9.31 EXAMPLES

Example 9.7

Consider a homogeneous, bare, spherical, source-free, critical, uranium-fueled reactor with an operating power of P. Under the proposed conditions, what would happen to the reactor power? Explain your reason. (a) This reactor is run at high power for a long time and (b) a group of visitors stands next to the reactor core.

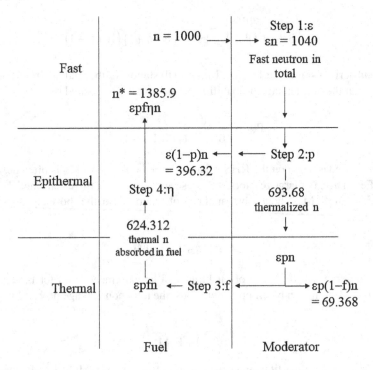

FIGURE 9.14 Life cycle of neutrons in the infinite core of $k_\infty = 1.3859$.

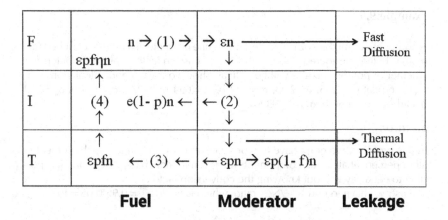

F		n → (1) →	→ εn	——————→	Fast Diffusion
	εpfηn		↓		
		↑	↓		
I		(4) e(1- p)n ←	← (2)		
		↑	↓		
		↑	↓ ———————		→ Thermal Diffusion
T		εpfn ← (3) ←	← εpn → εp(1- f)n		

Fuel   **Moderator**   **Leakage**

FIGURE 9.15 Four-factor formula with leakage for a thermal reactor neutron cycle.

Solution

a. For this situation, (1) fuel is being consumed and this implies that thermal fission factor, η, decreases; (2) fission products will be produced from fission reactions and this shows that the thermal utilization factor, f, decreases as the value in the denominator gets larger. Thus, k_{eff} will <u>decrease</u>.
   However, if Pu-239 is being bred from U-238 because this is the uranium-fueled reactor, this can offset k_{eff}. So, if this is not a breeder reactor, then situations (1) and (2) will be dominating and k_{eff} will decrease and eventually will be less than 1. Thus, the **power will drop**.

b. In this situation, the infinite multiplication factor still remains the same. However, this group of visitors acts as a neutron reflector or shielding causing the non-leakage probabilities to increase. This implies that there will be less leakage of neutrons and the reactor will become supercritical, implying that the **power will increase**.

Example 9.8

Show that when fissile material is being added to the core materials (fuel (F) + moderator (M)), the value of L^2 can be expressed in terms of $(L_M)^2$ of the moderator and thermal utilization factor, f.

Solution

We know that $L^2 \equiv D/\Sigma_a$. Since the main compositions will be mostly moderator (M). Thus,

$$L^2 = \frac{D^M}{\Sigma_a^F + \Sigma_a^M} = \frac{D^M}{\Sigma_a^M} \cdot \frac{\Sigma_a^M}{\Sigma_a^F + \Sigma_a^M} = L_M^2\left(1 - \frac{\Sigma_a^F}{\Sigma_a^F + \Sigma_a^M}\right) = L_M^2(1-f).$$

The proof is done.

Example 9.9

A homogenous mixture of 100 moles of graphite per mole of 5 mole% enriched uranium is being considered. Assume that the fast fission factor is about 1. Determine whether or not this mixture will go critical. Show your work and calculation.

Given data: U-235: $\nu=2.43$, $\sigma_f=582$ b, $\sigma_a=694$ b; U-238: $\sigma_s=8.3$ b, $\sigma_a=2.71$ b; carbon: $\sigma_s=4.66$ b, $\sigma_a=0.0034$ b.

Solution

In this problem, we don't have to worry about a conversion to weight percent or atom percent at all, as all components are given in mole basis. Therefore, it is possible to get away without knowing the conversion factor.

First, we have to find N^{238}/N^{235} ratio, which is $N^{238}/N^{235}=0.95/0.05=19$.

Next, $N^C/N^{235} = \dfrac{(100 \text{ mol C per mol U})}{(0.05 \text{ mol } {}^{235}U \text{ per mol U})} = 2000 \text{ mol C/mol}^{235}U.$

$\varepsilon=1$ and we need to determine p. Here,

$$I_{eff} = 3.9\left(\frac{\Sigma_s}{N^{238}}\right)^{0.415} = 3.9\left(\frac{\frac{N^U}{N^{235}}\sigma_s^U + \frac{N^C}{N^{235}}\sigma_s^C}{N^{238}/N^{235}}\right)^{0.415}$$

$$= 3.9\left[\frac{20(8.3)+2000(4.66)}{19}\right]^{0.415} = 51.389 .$$

Now, we calculate lethargy for both U-238 and C, which are $\xi_{U\text{-}238}=2/(238+2/3)=0.0084$ and $\xi_C=2/(12+2/3)=0.158$. So, the average lethargy can be calculated to be

$$\langle\xi\rangle = \frac{N^U\sigma_s^U\xi_U + N^C\sigma_s^C\xi_C}{N^U\sigma_s^U + N^C\sigma_s^C} = \frac{20(8.3)(0.0084)+2000(4.66)(0.158)}{20(8.3)+2000(4.66)} = 0.1554.$$

Then,

$$\rho = \exp\left(-\frac{N^{238}}{\xi\Sigma_a}I_{eff}\right) = \exp\left(-\frac{N^{238}/N^{235}}{\xi\Sigma_s/N^{235}}I_{eff}\right) = \exp\left(-\frac{19}{0.1554(9486)}(51.389)\right) = 0.5156. \text{ It}$$

should be noted that $(N^U\sigma_s^U + N^C\sigma_s^C)/N^{235}=9486$ b, which is the value of Σ_s.

Next, we have to determine the thermal utilization factor, which is

$$f = \frac{\sigma_a^{235} + \left(\frac{N^{238}}{N^{235}}\right)\sigma_a^{238}}{\sigma_a^{235} + \left(\frac{N^{238}}{N^{235}}\right)\sigma_a^{238} + \left(\frac{N^C}{N^{235}}\right)\sigma_a^C} = \frac{694+19(2.71)}{694+19(2.71)+2000(0.0034)} = 0.9909.$$

The last factor is the thermal fission factor.

$$\eta = v\frac{\Sigma_f^{235}}{\Sigma_a^U} = v\frac{\sigma_f^{235}}{\sigma_a^{235} + \dfrac{N^{238}}{N^{235}}\sigma_a^{238}} = \frac{(2.43)(582)}{694 + 19(2.71)} = 1.897$$

So, we can now calculate the infinite multiplication factor, which is

$$k_\infty = 1.0 \times 0.5156 \times 0.9909 \times 1.897 = \mathbf{0.969}.$$

Therefore, this homogeneous mixture material will make the reactor being **sub-critical** and we need to recast a ratio for a better k_∞, as it should be above 1 for a good design.

Example 9.10

Determine the infinite multiplication factor for a homogeneous mixture of ^{235}U and graphite with an atomic uranium to carbon ratio of 1 to 50,000? In this scenario, we are using pure ^{235}U diluting with carbon. Use Table 9.3 and the following data: U-235: $v = 2.43$, $\sigma_f = 582$ b, $\sigma_a = 694$ b; carbon: $\sigma_s = 4.66$ b, $\sigma_a = 0.0034$ b.

Solution

In this situation, we don't have ^{238}U in the mixture; thus, the value of εp will be about 1 and $k_\infty = \eta f$. From Table 9.3, $\eta = 2.07$ for pure ^{235}U. So,

$$f = \frac{\sigma_a^{235}}{\sigma_a^{235} + \left(\dfrac{N^C}{N^{235}}\right)\sigma_a^C} = \frac{694}{694 + 50000(0.0034)} = 0.8032.$$

Then, $k_\infty = 2.07 \times 0.8032 = \mathbf{1.662}$.

Example 9.11

Calculate the radius R of a critical bare sphere composed of a homogeneous mixture of ^{235}U and graphite with pure uranium-235 to carbon atom ratio of 1 to 50,000.

Solution

Now, we are combining the idea and look at the six-factor formula. For criticality, we will require to have

$$k_{eff} = \frac{\varepsilon p f \eta}{(B^2 L^2 + 1)}\exp(-B^2\tau) = 1.$$

From the previous example, we know that $\varepsilon p f \eta = 1.662$. And with this core mixture, we can determine L^2 for carbon to be $(59\,\text{cm})^2 = 3481\,\text{cm}^2$ (use the value from Table 9.5) and $\tau = 368\,\text{cm}^2$. So, the criticality condition is

$$k_{eff} = 1 = \frac{1.662}{\left(3481B^2 + 1\right)} \exp\left(-368B^2\right).$$

The solution for $B^2 = (\pi/R)^2$ can be obtained by trial and error. B is about 0.012735 cm^{-1} or $B^2 = 0.00016218$ cm^{-2} (see the plot below). From this result, we find the critical radius to be R = **246.68 cm** or roughly **247 cm**.

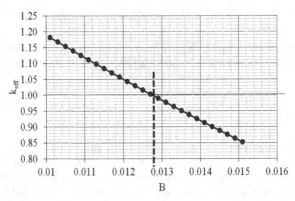

Important Terms (in the Order of Appearance)

Chain reaction	Infinite multiplication factor	Transport mean free path
Nuclear reactor	Homogeneous	Six-factor formula
Criticality	Change in reactivity	Five-factor formula
Leakage	Bare reactor	Four-factor formula
Poison	Non-leakage probability	Fast fission factor
Subcritical	Mixed oxide (MOX)	Resonance escape probability
Net leakage	Thermal reactor	Thermal utilization factor
Reflector	Fast reactor	Thermal fission factor
Supercritical	Moderator	Thermal diffusion area
Critical	Thermalization	Critical buckling
Critical mass	Enrichment	Thermal diffusion coefficient
Fuel	Most probable neutron velocity	Fermi age
Multiplication factor	Macroscopic slowing-down power	Migration length
Effective multiplication factor	Moderating ratio	Migration area
Excess neutron multiplication factor	Center-of-mass system	

BIBLIOGRAPHY

Bush, H.D., *Atomic and Nuclear Physics,* Prentice Hall, Englewood Cliffs, NJ, 1962.
Coombe, R.A., An Introduction to Radioactivity for Engineers. New York: Macmillan Publishing Co., Inc., 1968.
Duderstadt, J.J., and L.J. Hamilton, *Nuclear Reactor Analysis,* New York: Wiley, 1976.
El-Wakil, M.M, *Nuclear Power Engineering.* New York: McGraw-Hill Book Company, Inc., 1962.
Glasstone, S., and M.C. Edlund, *The Elements of Nuclear Reactor Theory*, D. Van Nostrand Company, New Jersey, 1952.

Glasstone, S., and A. Sesonske, *Nuclear Reactor Engineering*, 3rd ed., New York: Van Nostrand Reinhold, 1981

Glasstone, S., and R.H. Loveberg, *Controlled Thermonuclear Reactions*. Princeton, N.J.: D. Van Nostrand Company, 1960.

Lamarsh, J.R., and Baratta, A.J., *Introduction to Nuclear Engineering*, 3rd Ed., Prentice Hall, NJ, 2001.

Lewis, E.E., *Nuclear Reactor Physics*. New York: Elsevier, 2008.

Mayo, R. M., *Nuclear Concepts for Engineers*, American Nuclear Society, La Grange Park, IL, 1998.

Reactor Physics Division Annual Report, July 1, 1963 – June 30, 1964. ANL-7010 United States: Web.

Shultis, J.K. and R. E. Faw, Fundamentals of *Nuclear Science* and Engineering 2*nd* Ed., CRC Press Taylor & Francis Group, Boca Raton, FL, 2008.

FURTHER EXERCISES

A. True or False: If the statement is false, give a counterexample or explain the correction. If the statement is true, explain why it is true.

1. A good moderator should have a high moderating ratio and a low macroscopic slowing-down power.

2. The cross section at the average neutron velocity should increases as temperature increases.

3. At the most probable velocity of neutron, the kinetic energy is ~0.025 eV.

4. The ratio of the average velocity to the most probable velocity of neutron is a function of temperature.

5. An infinite multiplication factor is dependent of the reactor core's geometry.

6. The thermal fission factor is the number of fast neutrons produced per thermal neutron absorbed in the fuel.

7. Ten mole percent-enriched uranium implies that the mole ratio between U-235 and U-238 is 9.

8. By keeping the same fuel amount but increasing the moderator quantity, the thermal utilization factor should remain the same.

9. For a real core to be critical ($k_{eff} = 1$), infinite multiplication factor must be larger than one to compensate for neutron leakage, buildup of fission fragments, consumption of fissionable nuclei, and changes in temperature and pressure in the core.

10. By increasing the amount of graphite in the reactor core, the lethargy value of graphite should increase as well.

B. Problems

1. Verify that neutrons of speed 2200 m/s have an energy of 0.0253 eV: (a) using the KE aspect and (b) using the Boltzmann constant aspect. The rest mass of a neutron is 1.675×10^{-27} kg and assuming 300 K for room temperature.

2. If the neutron absorption cross section of boron at 0.0253 eV is 760 barns, what would it be at 0.1 eV?

3. Calculate the microscopic absorption cross section in barns of the element zirconium using the isotopic data in the following table.

Mass Number	Abundance (atom %)	Absorption Cross Section (Barns)
90	51.45	0.014
91	11.22	1.2
92	17.15	0.2
94	17.38	0.049
96	2.80	0.020

4. (a) Calculate the macroscopic cross section for scattering of 1 eV neutrons in water, assuming an atom number density for water of $0.0334 \times 10^{24} cm^{-3}$ and scattering cross sections of 20 barns for hydrogen and 3.8 barns for oxygen. (Assume that the microscopic cross sections for different reaction types are additive.) (b) Find the scattering mean free path λ_s in cm.

5. A thermal neutron interacts with a ^{235}U nucleus. What is the probability that the neutron will undergo a scattering interaction assuming the following microscopic cross-section values: $\sigma_t = 700$ b, $\sigma_\gamma = 99.3$ b, $\sigma_f = 587$ b?

6. Consider the following microscopic thermal cross sections for $^7Li_3$: $\sigma_a = 0.05$ b (absorption) $\sigma_s = 1.04$ b (scattering) $\sigma_t = 1.09$ b (total).

    What is the probability that if a thermal neutron interacts with a $^7Li_3$ nucleus, it is absorbed?

7. What is the mean free path in cm of a thermal neutron in natural thorium assuming a total microscopic cross section of 20.4 b and a density of 11.72 g/cm^3?

8. The effective multiplication factor of a reactor is 0.94. (a) Calculate the reactivity of the reactor. (b) Determine the change in core reactivity in order to make the reactor critical?

9. A parameter termed the Doppler fuel temperature coefficient α_F provides the change in core reactivity $\Delta\rho$ that occurs as a result of a change in the core average fuel temperature ΔT_F. That is, $\alpha_F = \Delta_\rho / \Delta T_F$

    Reactors are typically designed to be operated with a negative temperature reactivity coefficient so that a rise in fuel temperature that occurs as a result of an increase in core power tends to reduce the associated increase in core reactivity. Similarly, a reduction in fuel temperature will tend to increase the core reactivity. Assume the temperature reactivity coefficient for a reactor is $-1.1 \times 10^{-5}/°C$.

    a. If the core reactivity of a reactor is -0.0005, how many degrees Celsius must the fuel temperature be changed to make the reactor critical?

    b. Is this temperature change an increase or decrease in the fuel temperature?

10. The macroscopic fission cross section of an *infinite*, homogeneous reactor has a value of 0.08 cm^{-1}. On average, 2.5 neutrons are produced per fission. What is the macroscopic absorption cross section of the reactor in cm^{-1} if the reactor is *critical*?

Data for Problems 11 and 12: assume the following *fast* neutron group constants.

Material	v	σ_f (b)	σ_γ (b)	σ_t(b)
^{235}U	2.6	1.4	0.25	6.8
^{238}U	2.6	0.095	0.16	6.9
^{239}Pu	2.98	1.85	0.26	6.8
Na (Sodium)	–	–	0.0008	3.3

11. Assume that an *infinite fast* reactor is composed entirely of a homogeneous mixture of ^{239}Pu and sodium. Using the properties for ^{239}Pu and sodium given in the above table, on the average how many sodium atoms must there be per ^{239}Pu atom in order to make the reactor critical?

12. A *fast bare* reactor is composed entirely of ^{235}U and ^{238}U. If the number density of ^{238}U in the core is four times that of ^{235}U, what is the value of k_∞ for the reactor?

13. A proposed reactor core is considered to have a homogeneous mixture of ^{235}U and graphite with an atomic uranium to carbon ratio of 1 to 10,000. Calculate (a) k_∞ and (b) the radius R of a critical bare sphere. In this scenario, we are using pure ^{235}U diluting with carbon. Use Table 9.3 and the following data: U-235: $v = 2.43$, $\sigma_f = 582$ b, $\sigma_a = 694$ b; carbon: $\sigma_s = 4.66$ b, $\sigma_a = 0.0034$ b.

14. A bare, spherical uranium-fueled reactor is operating at a power P for a critical operation. This is a source-free reactor. Discuss how and why the power increases, decreases, or remains unchanged as a result of each of the separate events to the reactor.
   a. Deforming the shape of the reactor into a cube.
   b. Having a person right next to the core.
   c. Launching a reactor core into outer space.
   d. Wrapping a thin sheet of cadmium around the core.
   e. Introducing Pu-239 into the core.

NOTE

1 We will explore the general concepts first and then provide basic mathematical physics associated with these concepts to help us understand on how to control the nuclear reactors.

10 Nuclear Reactors, Power, and Fuel Cycles

OBJECTIVES

After studying this chapter, the reader should be able to:

1. Know general features of nuclear power plants, different nuclear reactor types, and generation I–IV designs.
2. Understand the basic design of the pressurized water reactor and its reactor physics, operation, and controls.
3. Grasp basic nuclear fuel cycle concepts and brief processes associated with front-end and back-end stages.

10.1 GENERAL FEATURES OF NUCLEAR POWER PLANTS

A simple schematic of the design of the reactor part of a nuclear power plant is shown in Figure 10.1. All reactors have a core where the fuel is housed and the neutron-induced fission takes place. Should the reactor be a solid-fuel fast breeder reactor, the core would typically be surrounded by a blanket region initially loaded with fertile material and where the bulk of the breeding would occur. In this case, the fuel in the core is often termed the driver fuel.

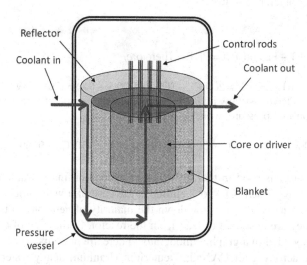

FIGURE 10.1 Principal components of the reactor part of a nuclear power plant.

DOI: 10.1201/9781003272588-10

In thermal reactors, the core is surrounded by a reflector region containing a moderating material whose function is to reduce the net neutron leakage of the core by scattering many of the escaped neutrons back into the core.

The primary means of controlling the reactor power is typically achieved by the movement in and out of the core of control rods composed of material with high neutron absorption properties. The required components are housed inside a reactor vessel. In the case of a **light water reactor** (LWR), this reactor vessel is termed the **reactor pressure vessel** since the coolant is maintained at a high pressure in order to increase the thermodynamic efficiency of the plant and limit the amount of boiling of the water coolant. The heat generated by the reactor is removed by the **coolant**. The coolant typically enters the reactor vessel at an elevation above the top of the core, flows down around the outside of the core, and then enters the core from the bottom. The coolant captures the heat generated by the fission reactions as it moves upward through the core, exiting the reactor vessel typically at the same elevation at which it entered. In a water-cooled reactor, the coolant flowing down around the outside of the core also serves as a reflector.

Basically, **thermal conversion efficiency** (TCE) can be expressed as the ratio of the electrical power (PE) to the actual thermal power (PT), that is, TCE = PE/PT. Based on this information, rejected heat rate (PR) of the system can be calculated by PR = PT − PE. For example, 100 MW(e) with thermal efficiency of 25% would have 400 MW(t) and the heat rejection rate would be 400 − 100 = 300 MW.

Example 10.1

A 2000 MWt nuclear power plant has a thermal conversion efficiency of 40%. (a) How much thermal power (MW) is rejected through the condenser to cooling tower? (b) What is the flow rate (kg/s) of the condenser cooling water if the positive temperature difference is 30°C?

Solution

Here, PR = PT − PE and TCE = PE/PT. Therefore,

a. PR = PT(1 − PE/PT) = 2000(1 − TCE) = 2000(1 − 0.4) = 1200 MW.
b. Here, PR = (mass flow rate) × C_p × (ΔT), where C_p = specific heat of water at constant pressure, which is 4180 J $kg^{-1}C^{-1}$.

So, Mass flow rate = (1200 × 106 J/s)/[(4180 J/kg/°C) × (30°C)] = 9569 kg/s.

Coolant materials used in thermal reactors include ordinary water (also termed light water), heavy water, CO_2, or helium. In the case of a water coolant, the water also serves as a moderator to slow down the majority of neutrons to thermal energies before they are absorbed or leak from the reactor. In the case of a gas coolant, the fuel is embedded in a graphite moderator. There are of course exceptions. In the **CANDU** (the acronym for CANada Deuterium Uranium) heavy water reactor, the fuel assemblies are housed inside individual pressure tubes through which the heavy water flows, thereby removing the need to house the reactor core components inside a pressure vessel. The obsolete Soviet **RBMK** (the acronym for Reaktor Bolshoy Moshchnosti Kanalnyy, which represents "high-power channel-type reactor") design also used a pressure tube arrangement, but with light water coolant. Because of the uniqueness of the reactor physics for the RBMK design, housing the pressure tubes inside graphite blocks was necessary to provide sufficient neutron moderation.

Typical coolants for a fast reactor are sodium or lead in liquid (molten) phase or helium. Since fast reactors operate at relatively low coolant pressures, reactor vessels able to withstand high pressures such as those used for LWRs are no longer a requirement.

The proposed design of the **molten salt reactor** (MSR), a thermal breeder based on the thorium fuel cycle, has the fuel dissolved in a molten salt coolant and therefore has little resemblance to the reactor design depicted in the above schematic.

Most existing reactor designs produce electricity by a steam-driven turbo-generator. The reactor along with the associated steam-producing components are divided into two parts. The systems and components inside the containment building are termed the **nuclear steam supply system** (NSSS). The remaining systems outside the containment building are termed the **balance of plant**. A simplified depiction of a single-loop **pressurized water reactor** (PWR) is shown in the Figure 10.2 and will be discussed in details later.

With the exception of the **boiling water reactor** (BWR), reactors with liquid coolants usually have a heat exchanger as the interface between the primary coolant

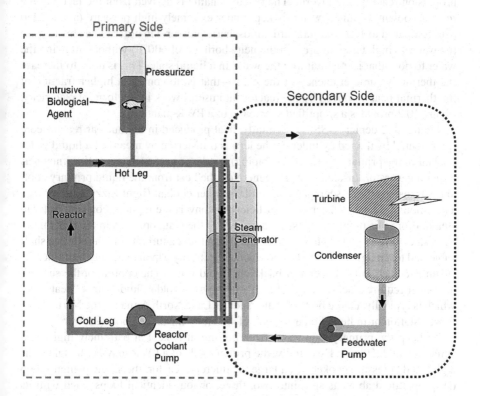

FIGURE 10.2   Simplified schematic of a single-loop PWR.

that removes heat from the reactor core and the secondary liquid water coolant which is boiled to steam to turn the turbine by absorbing the heat from the primary coolant. Such heat exchangers are typically referred to as steam generators. The final major component of a commercial nuclear power plant is a *containment* structure inside of which is housed in the NSSS.

In the next few sections, we will discuss common types of nuclear power reactors and focused the operation on PWR for a basic understanding of a system control.

10.2   LIGHT WATER REACTORS

As discussed briefly in the first frame, most reactors are moderated, reflected, and cooled by ordinary (*light*) water. Water has excellent moderating properties. Water has a high vapor pressure, which means that LWRs must be operated at high pressures. Enriched uranium is required to fuel LWRs because water absorbs thermal neutrons. There are two common types: (1) Pressurized Water Reactor (PWR) and (2) Boiling Water Reactor (BWR).

10.3   PRESSURIZED WATER REACTOR

The PWR is by far the most common type of commercial power reactor and naval propulsion reactor in the world. The reactor's name is derived from the fact that the reactor coolant, ordinary water, is kept under extremely high pressure (about 2250 psia compared to 14.7 psia standard atmospheric pressure) in order to be able to raise the water to high temperatures (in the neighborhood of 600°F) without allowing the water to boil, that is, maintaining the water in a liquid state. This is done to increase the thermodynamic efficiency of the plant—that is, to convert a higher fraction of the thermal heat produced by the reactor into useful work in the form of electricity. Figure 10.2 provides a simplified schematic of a PWR plant.

Figure 10.2 depicts NSSS. Essentially, heat produced in the nuclear reactor core (represented by the red cylinder on the left) is transferred by means of a liquid water coolant in the primary system (blue piping) to a large heat exchanger, also known as a steam generator. Inside the steam generator, the heat from the liquid primary coolant is transferred to a segregated secondary water coolant (light/fuzzy blue piping). The amount of heat transferred is sufficient to convert the liquid secondary water to steam. From this point on, the secondary side of the plant operates in the same manner as a coal-fired power plant; that is, the steam turns a turbine which rotates a shaft subjected to a magnetic field, thus producing electricity. Upon exiting the turbine, the steam is cooled in the condenser back into liquid water. The cooling of the secondary water requires the exchange of heat with an even colder fluid from a "heat sink" which is typically a large body of water such as Lake North Anna for the North Anna Power Station or in lieu of a large body of water, by means of cooling towers.

To keep the primary side water from boiling, it is kept at extremely high pressure—about 2250 psia. Essentially, the primary side of a PWR may be thought of as a glorified "pressure cooker." In a large commercial reactor, the single primary loop design pictured above is split into two, three, or four identical loops, each with its own steam generator as depicted in Figure 10.3.

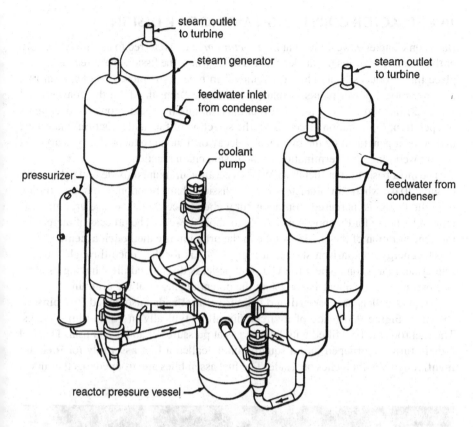

steam outlet
to turbine

steam generator

feedwater inlet
from condenser

coolant
pump

pressurizer

steam outlet
to turbine

feedwater from
condenser

reactor pressure vessel

FIGURE 10.3   Four-loop PWR primary system.

To maintain the primary system coolant at a constant target pressure, one of the "hot leg" loops is attached a *pressurizer*. In Figure 10.2, the pressurizer is the component containing the "intrusive biological agent." Just to be clear, in a real reactor no form of life more advanced than possibly certain forms of bacteria could survive in the primary system environment of a PWR; that is, no fish were harmed in the creation of Figure 10.2.

The pressurizer contains both liquid water and steam. Should the pressure of the primary system increase above the target value, cold water is sprayed into the top of the pressurizer, thus condensing some of the steam and reducing the pressure. On the other hand, should the primary system pressure start to decrease below the target value, heaters located at the bottom of the pressurizer activate, increasing the water temperature and raising the pressure back to the target value. The pressurizer is important in preventing the primary system from going "solid," that is, being completely filled with liquid water. This would create a hazardous condition, since liquid water is pretty much an incompressible fluid. Under such a condition, a small increase in coolant pressure would result in significant pressure stresses being exerted internally on the primary system piping.

10.4   REACTOR CORE DESIGN AIDING PWR DESIGN

Based on Chapter 9, we know that the *reactor core*, also referred to as simply the *core* or the *reactor*, contains the nuclear fuel. This is where the fission chain reaction takes place that produces heat. The heat removal in reactors is more complex than that in conventional power plants in that (1) fuel is not "consumed" in the conventional sense; (2) in solid-fuel reactors, the fuel must maintain a fixed geometry over years of operation; (3) radiation effects limit the selection of fuel and structural materials; and (4) heat generated by the radioactive decay of fission products (decay heat) must be removed long *after* termination of the fission chain reaction.

The most basic unit of fuel in a PWR is a ceramic uranium dioxide fuel pellet, UO_2 (Figure 10.4), where, in order to sustain a fission chain reaction, the concentration of fissile U-235 is increased from its natural abundance of 0.7% of the uranium (the remainder being for the most part U-238) to about 3%–5%. The process of increasing the concentration of the U-235 isotope in the uranium is called **enrichment**.

The energy comparison shown in Figure 10.5 is for the "once-through" nuclear fuel cycle; that is, one where the UO_2 fuel pellet is discharged after having used up approximately only about 1% or 2% of the potential energy of the uranium.

The fuel pellets are inserted into cylindrical **fuel rods** (also called **fuel pins**, as shown in Figure 9.3) made of a zirconium alloy, typically in 12-foot-high stacks. The fuel rods are backfilled with helium gas at pressures of about 200 psia. The fuel rods in turn are grouped into a square bundle called a **fuel assembly** (or **fuel element**) roughly eight inches on a side. The fuel assemblies are grouped together into a

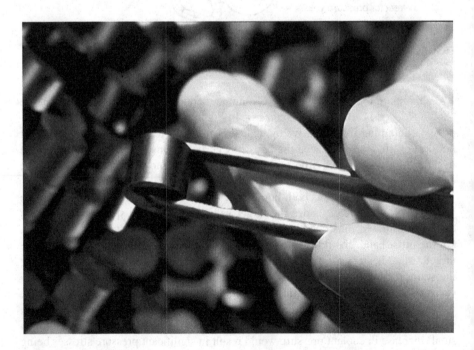

FIGURE 10.4 UO_2 fuel pellet.

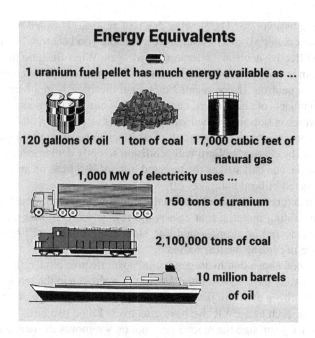

FIGURE 10.5 Energy equivalents for a single UO$_2$ fuel pellet.

reactor core of roughly cylindrical shape. PWR assembled reactor core. The depicted core would consist of about 200 fuel assemblies producing about 3200 MW of thermal heat which in turn would generate about 1000 MW of electricity.

10.5 BACK TO REACTOR PHYSICS TO HELP UNDERSTANDING PWR DESIGN

The heat inside a reactor core is produced as a result of neutron-induced nuclear fission as depicted in Figure 7.5. The "magic bullets" that cause certain atomic nuclei to fission are neutrons. A chain reaction is possible since a fission induced by a single neutron results on average in the creation of roughly 2.5 new neutrons. These neutrons are born at high energies and as such are termed fast neutrons.

Those nuclides most susceptible to being fissioned are termed fissile nuclei of which U-235 is the only one to occur naturally. It turns out that the probability of a neutron causing fission depends not only on the target nucleus, but also on the neutron's energy or speed. The probability of low-energy neutrons causing a fissile nuclide to fission is several hundred times that of a fast neutron. Such low-energy neutrons are termed *thermal neutrons* and reactors designed to take advantage of such slow, low-energy neutrons are termed "thermal" reactors. PWRs are thermal reactors.

The new neutrons being created from nuclear fission are fast neutrons. So how do we cause the fast neutrons to lose enough energy to become thermal neutrons in order to increase the fission rate? The slowing down of neutrons is called moderation, and it turns out that materials of low atomic weight make good neutron "moderators."

The most effective neutron moderator is hydrogen. Why? Because the nucleus of a hydrogen atom is composed of a single proton. The mass of a proton is nearly identical to that of a neutron. If you recall Newtonian mechanics, in a head-on collision between two objects of equal mass, one of which is moving (the neutron in our case) and one of which is stationary (the hydrogen nucleus or proton in our case), it is possible for the moving object to impart all of its kinetic energy to the stationary target object. This is the classic "billiard ball" collision so aptly demonstrated in the game of pool when the cue ball strikes the targeted billiard ball head on and thereby stops while the targeted billiard ball takes off.

Thermal reactors are designed around this concept—that is, they all contain some effective moderating material that causes the fast neutrons to lose energy through atomic collisions and be transformed into slow neutrons that will be significantly more effective in causing nuclear fission.

We mentioned above that hydrogen is the most effective moderating material. In a PWR, we already have a substantial amount of hydrogen in the reactor core in the form of the hydrogen atoms in the water coolant. So, in the design of a water-cooled thermal reactor such as a PWR, we have effectively killed two birds with one stone. The water flowing through the reactor core not only removes the heat but while during so it slows down the fast fission neutrons to thermal energies, thus significantly increasing the rate of fission chain reactions.

10.6   REACTOR OPERATION FOR PWRs

Reactor power is changed by changing the size of the neutron population inside the reactor core. A growing neutron population means an increase in the number of fission reactions and thus an increase in the reactor power. Conversely, a declining neutron population results in a decrease of the reactor power. Reactor operators change the power in a PWR by one of the following two means:

1. The insertion into the core or withdrawal from the core of rods containing effective neutron absorbing materials. These are termed **control rods**.
2. Changing the concentration of a neutron absorbing material dissolved in the reactor coolant. In a PWR, boric acid is such a material. It is added to or removed from the coolant water since boron-10 is a good absorber of thermal neutrons.

In the event of an emergency, the fission chain reaction is quickly terminated by a **reactor trip** or SCRAM. This is accomplished by inserting into the core a large number of control rods. This is shown through Figure 10.6.

A key challenge in the design of nuclear fission reactors is that tripping the reactor doesn't stop the reactor core from producing heat. Remember those fission fragments discussed previously? They are all radioactive, so even if the fission chain reaction is terminated, the fission fragments and their daughter radioisotopes, the fission products, will continue to radioactively decay. Radioactive decay produces

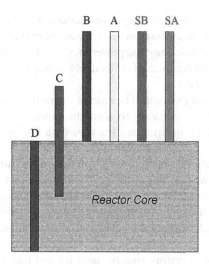

FIGURE 10.6   Simplified depiction of the insertion of control rods—the brown D-bank rod being fully inserted, the green C-bank control rod being partially inserted, and the remaining control rods being completely withdrawn from the core.

heat. This is termed **decay heat**. When operating, about 7% of the heat being produced in the reactor core is not directly from the fission reactions but from the fission product decay heat. In a large commercial reactor, unless this decay heat is continually removed from the reactor core, enough of it will build up so as to damage the fuel. This is what caused the problems at both the Three Mile Island accident and the Fukushima event—the failure to remove the decay heat.

In most of the older operating reactors, pumps are necessary to keep the coolant moving through the reactor core in order to remove the decay heat. Pumps require power to run. At Fukushima, the combined effects of the earthquake and tsunami wiped out all sources of power needed to operate the pumps. Although the reactors tripped and thereby were no longer maintaining fission chain reactions, the lack of pumping power resulted in the decay heat increasing the temperature of the reactor fuel to the point where it was damaged, thus releasing some of its radioactive inventory into the coolant.

Many of the reactors under construction and those on the drawing boards are designed such that no electrical power is required to remove the decay heat following a reactor trip. Instead, the decay heat is removed by passive systems such as the natural circulation of the coolant through the reactor core.

10.7   REACTIVITY FEEDBACK FOR PWRs

Finally, we come to the most abstract of the concepts to be presented, which is negative **reactivity feedback**. This is important in understanding reactor operation. In this case, we will focus on one important type of such feedback—the moderator (or coolant) feedback. Essentially negative moderator feedback means that as the coolant heats up due to an increase in core power, there is a natural mechanism that comes into play that tends to limit the power increase. Conversely, during a decrease in core power, the negative moderator feedback tends to slow down and limit the amount of power decrease.

Let us illustrate by an example. The reactor operators wish to increase the reactor power by a couple of percent. To start the power increase, they withdrawal some of the control rods inserted into the core by a few inches. This means there is now less-neutron absorbing material in the core, so the neutron population starts to increase. This in turn increases the neutron fission reaction rate and thereby the core power. If there was no negative feedback mechanism in effect, the reactor power would continue to increase until either the fuel melted or the reactor operators inserted some control rods further into the core, thereby increasing the rate of neutron absorption and bringing the core neutron population back to a steady-state condition.

Instead, the reactor operators simply stand by and take no action. After a few minutes, the increasing reactor power levels off and reaches a new steady-state value. Why? Well, there are a couple of important negative feedback mechanisms taking place. One called the Doppler or fuel temperature feedback we won't describe in any detail here, but it is a factor in all reactor designs. Instead, we'll focus on the moderator feedback mechanism.

The increase in reactor power causes the temperature of the coolant water in the core to increase. What happens to the density of liquid water when you heat it? The water expands so the density decreases. This means the number density of the hydrogen nuclei in the core decreases. There are now fewer hydrogen nuclei in the core to moderate the fast fission neutrons, which means that fewer fission neutrons slow down to thermal energies. Fewer thermal neutrons means that the fission chain reaction rate decreases. Eventually, the reduction in the fission rate stops the power increase, and a new equilibrium (steady-state) condition is reached.

The core of the Soviet reactor which was destroyed in the Chernobyl accident had a *positive* reactivity feedback, so when the power in the core began to increase, the positive feedback accelerated the rate of power increase. Unlike the Soviet reactors, all U.S. plants are designed to be operated with a negative reactivity feedback.

10.8   BOILING WATER REACTORS

It was thought that if water were permitted to boil within a reactor, dangerous instabilities would result because of uneven formation and movement of the steam bubbles. This is true at low-pressure operation. However, boiling is stable under higher pressure operation. Advantage of BWR is that steam is formed in the reactor itself and goes directly to the turbine (see Figure 10.7). Thus, there is no need for the steam generators like in PWR. Basically, BWR is the direct cycle. It requires less water to be pumped through a BWR per unit time than PWR for the same power output. However, water becomes radioactive while passing through the reactor core. All of the components of the steam utilization system, the turbines, condenser, re-heaters, pumps, piping, and so on, must be shielded in a BWR plant.

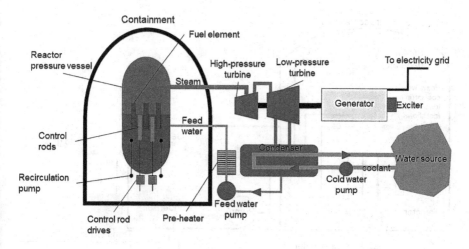

FIGURE 10.7   A schematic of BWR direct.

It has a maximum pressure about 7 MPa and a thinner pressure vessel wall than a PWR.

However, power density (watt/cm³) is smaller in a BWR than in a PWR; thus, overall dimensions of a pressure vessel for a BWR are larger than for a PWR of the same power. Fuel for a BWR is essentially the same as for PWR. BWR control rods are always placed at the bottom of the reactor rather than at the top as in the case of PWR. It can provide thermal energy efficiencies of about 34%, which is similar to the PWR.

10.9   HEAVY WATER REACTOR (HWR)

This has been under development in several countries, especially Canada. This interest dates back to WWII when the country was deeply involved in HWR research for military purposes. HWR can be fueled by *natural uranium*. This is quite an advantage as there is no requirement for the construction of the uranium enrichment plants. The Canadian HWR, known as **CANDU** has been successful in Canada and has been exported abroad (see Figure 10.8).

Absorption cross section of deuterium for thermal neutrons is very small—much smaller, for example, than the x-section of ordinary hydrogen. However, D in D_2O is twice as heavy as H in H_2O. D_2O is not as effective in moderating neutrons as H_2O. Neutrons, on the average, lose less energy per collision in D_2O than they do in H_2O, and they require more collisions and travel greater distances before reaching thermal energies than in H_2O. From this aspect, HWR core is larger than that of an LWR. To avoid a large pressure vessel, CANDU utilizes the pressure tube concept. The thermodynamic efficiency of CANDU plants is only between 28% and 30%. It is possible to use enriched fuel as well (~1.2 w/o appears to be optimum), but that is just optional from the original goal.

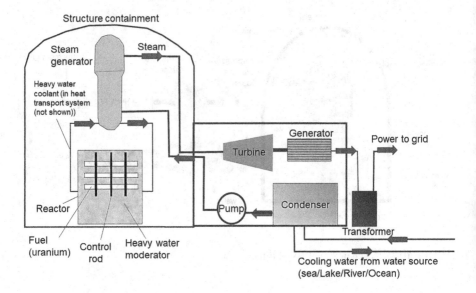

FIGURE 10.8   CANDU reactor schematic.

10.10   GAS-COOLED REACTOR (GCR)

Natural uranium, graphite-moderated reactors were developed in the United States during WWII to provide conversion of U-238 to Pu-239 for military use. In this general reactor, coolant gas is CO_2 (see Figure 10.9). This gas is not a strong absorber of thermal neutrons and it does not become excessively radioactive. At the same time, CO_2, is chemically stable below 540°C and does not react with either the moderator or fuel. One of the great advantages of GCRs is their high thermal efficiency of 40%.

In the US, GCR technology has been centered at the General Atomic Company, the **high-temperature gas-cooled reactor (HTGR)**. HTGR is a graphite-moderated, helium-cooled thermal reactor. Helium is an excellent coolant and more inert

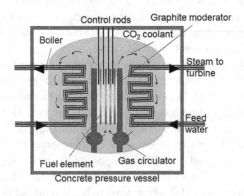

FIGURE 10.9   Schematic of gas-cooled reactor.

than carbon dioxide. It does not absorb neutrons, and therefore does not become radioactive. However, helium picks up small amounts of radioactive gases which escape from the fuel along with radioactive particles from the walls of the cooling channels. Thus, the coolant is still radioactive. Fuel is composed of thorium and highly enriched uranium. In time, the U-233 converted from the thorium replaces some of the U-235. The reactor does not breed (i.e., the conversion is less than unity, and this concept will be discussed in the nuclear power plant course); therefore, U-235 must always be present. The capital cost of HTGR is much less because the core contains U-235, fertile Th-232, and all the recycled U-233 (high enrichment). Its significant outcome is an achievement of its efficiency as high as 50%.

Now, we are going to explore the breeder reactor types.

10.11   BREEDER REACTORS

It is important to acknowledge that U-235 (even reserved) will not be enough to support the growing nuclear power industry. Thus, it is important to introduce breeder reactors. Here, the fuel base will be switched from U-235 to U-238 or thorium, both are considerably more plentiful than U-235. In addition, all of the depleted uranium (the residual uranium, mostly U-238) remaining after the isotope enrichment process can be utilized as breeder fuel. The need for the breeder is less in doubt. Great Britain, France, the Federal Republic of Germany, and Japan have no indigenous uranium resources and want to fully utilized all the imported uranium. There are four types of breeder reactors existing to date: (1) Liquid Metal-Cooled Fast Breeder Reactor (LMFBR), (2) Gas-Cooled Fast Breeder Reactor (GCFR), (3) Molten-Salt Breeder Reactor (MSBR), and (4) Light-Water Breeder Reactor.

10.12   LIQUID METAL-COOLED FAST BREEDER REACTOR

It was discovered before the end of WWII. The first Experimental Breeder Reactor (EBR) was a small plutonium-fueled, mercury-cooled device, operating at 25 kW, first went critical in 1946 in Los Alamos, New Mexico. A 1.3 MW breeder cooling with a mixture of sodium and potassium, was placed in operation in 1951 at ANL-West in Idaho, known as EBR-I. The system produces 200 kW of electricity; the world's first nuclear-generated electricity came from LMFBR. Chain reaction is maintained by fast neutrons, and there is no moderator in the core. Core and blanket contain only fuel rods and coolant. Liquid metals have excellent heat transfer characteristics and do not require pressurization to avoid boiling. However, sodium becomes radioactive and reacts violently with water. Here, U-238 converted to fissile Pu-239 or Th-232 into fissile U-233. The overall plant efficiency is about 40%.

10.13   GAS-COOLED FAST BREEDER REACTOR

Its model is an extension of the HTGR technology. It is a helium-cooled, reactor, fueled with a mixture of plutonium and uranium. The GCFR, unlike the LMFBR, requires no intermediate heat exchangers. Since the coolant in GCFR does not become overly radioactive, it is possible to work on any part of the coolant loops soon after the reactor is shut down. Some authorities believe that the GCFR is potentially capable of providing the lowest power costs of any reactor conceived to date.

10.14 MOLTEN SALT BREEDER REACTOR

This is a thermal breeder using U-233-thorium cycle. Here, U-233 is the only fissile isotope capable of breeding in a thermal reactor. Fuel, fertile material, and coolant are mixed together in one homogeneous fluid composing of fluoride salts ($72\,mol\%$ ^7LiF, $16\,mol\%$ BeF_2, $12\,mol\%$ ThF_4, and $0.3\,mol\%$ $^{233}UF_4$). There are several advantages of MSBR; these are as follows: (1) it requires very small fuel inventory due to good neutron economy (about 1.0–1.2 kg of fissile material per MW(e) of plant output compared to about 3 kg per MWe for an LWR or 3–4 kg per MW(e) for LMFBR), (2) it operates at just a little above atmospheric pressure because of low vapor pressure of the molten salts (thus, no expensive pressure vessel is required), and (3) it can produce superheated steam at 24 MPa and 540°C since high temperatures are possible with the molten salts, which leads to a very high overall plant efficiency of about 44%.

However, there are several disadvantages of MSBR as well. First, as the fuel flows of the reactor, it carries the delayed neutron precursors with it, and when the neutrons are emitted, they activate the entire fuel-containing loop. Second, maintenance and repairs to any component in the system may therefore require extensive reactor downtime and extensive automatic, remotely operated equipment. Third, the breeding ratio of the MSBR is in the 1.05–1.07 range, much smaller than for the LMFBR or GCFR. Lastly, doubling times are expected to be from 13 to 20 years. These last two points will be discussed further in the nuclear power plant course.

Now, let's us turn the table slightly to discuss about the concept behind "generation."

10.15 EVOLUTION OF NUCLEAR POWER REACTORS

During the 1950s, the first-generation (also known as **Gen I**) nuclear power reactors were designed, built, and operated to demonstrate the feasibility of commercial nuclear-generated electric power. These first-generation plants were followed by the second generation, often referred to as **Gen II** or Generation II reactors. Most of the world's nuclear power is generated by Gen II reactors built, for the most part, in the 1970s and 1980s, with design lives of 30 –40 years. Based on data and operating experience since gained, the design lives of many of the Gen II reactors have been extended by regulatory authorities to 50 or 60 years. By 2013, 75% of the commercial Gen II reactors in the U.S. had been granted license extensions to 60 years.

Reactors being built today are based on designs originating in the 1990s and are termed **Gen (Generation) III** reactors. They incorporate two different designs: (1) *Evolutionary* designs being based on extensions of existing Gen II designs stressing safety, efficiency, and standardization. Examples are the advanced boiling water reactor (ABWR) and System 80+. (2) *Passive* designs stressing passive safety features involving gravity, natural circulation, and pressurized gas as driving forces for cooling and residual heat removal as in the AP600 (600 MWe PWR designed by Westinghouse Electric Company), AP1000 (evolutionary improvement of AP600 by Westinghouse Electric Company) and ESBWR (Economic Simplified Boiling Water Reactor, designed to surpass its predecessor ABWR).

In 2001, the U.S. and nine other countries agreed to develop Generation IV or **Gen IV**, which are the designed model that be deployed after 2030 and afterward. It must operate at higher temperature range, 510–1000°C as opposed to Gen II and Gen III (~330°C). There are several enhanced features including (1) capital cost reduction, (2) advanced passive safety systems, (3) incorporate nuclear fuel cycle minimizing waste products, and (4) nonproliferation/safeguard advancement. The output range is expected from 150 to 1500 MW(e). As of 2012, the international forum has selected six designs, which are: (1) Very High-Temperature Reactor—graphite-moderated, He-Cooled; (2) Supercritical Water-cooled Reactor—high-temperature, high-pressure, water-cooled, operating above the thermodynamic critical point of water; (3) Gas-Cooled Fast Reactor—He-cooled; (4) Lead-Cooled Fast Reactor (LFR)—Pb–Bi eutectic mixture as coolant; (5) Sodium-Cooled Fast Reactor—may be used to burn actinides; and (6) MSR—epithermal neutron spectrum, circulating liquid mixture of uranium and a molten salt as fuel.

10.16   WHAT IS THE NUCLEAR FUEL CYCLE?

Various activities associated with the production of electricity from nuclear reactors are referred to collectively as the **"nuclear fuel cycle (NFC)."** These *activities* change and vary from country to country. A cartoonish sketch of the nuclear fuel cycle is shown in Figure 10.10. We can see that NFC provides important concepts and shows industrial scales of inputs (e.g., uranium ore, chemical processes, fabrication of materials), processes, and outputs (e.g., energy, important actinides, wastes).

We can breakdown Figure 10.10 into detailed processes, as shown in Figure 10.11, showing front-end and back-end routines. The process and its objective are listed in Table 10.1 with the color scheme associated to Figure 10.11.

Without process #8 in Figure 10.11, this can be referred to as **"open or once-through"** **fuel cycle**. Used fuel discharged from the reactor is treated as waste and fuel is kept in at the reactor pool. Waste is kept in interim storage before being disposed of in geological disposal. The U.S. has adopted this type, as other countries.

Both Figures 10.10 and 10.11 represent **"closed"** **fuel cycle**. This means *recycling*. Here, used fuel is reprocessed to extract U and Pu from FPs and other actinides. We can reuse of recovered fissile uranium in uranium oxide (UOX) or mixed oxide (MOX). Then, we will use of Pu in MOX fuel and burn MOX in LWR or FBR. Currently, France is one of the leading nuclear energy nations that has adopted a closed fuel cycle.

10.17   FRONT-END STAGES

The important objective of the front end is to *fabricate nuclear fuel*. In **mining and milling**, we utilize the fact that uranium has average crustal abundance 2–5 ppm,

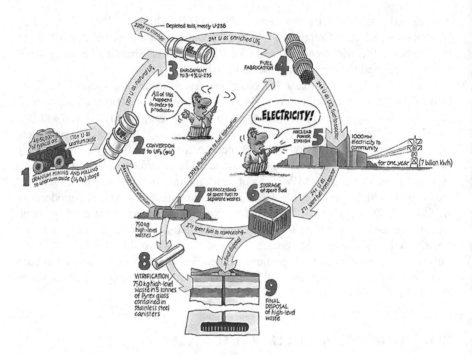

FIGURE 10.10 Schematic diagram of the nuclear fuel cycle.

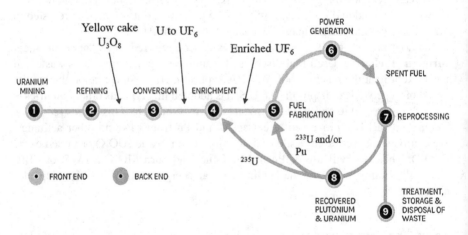

FIGURE 10.11 Schematic diagram of NFC showing front-end and back-end processes.

TABLE 10.1

Front-End and Back-End Processes of NFC

Process	Objective
Mining and milling	Uranium extraction
Conversion and enrichment	Fuel material via UF_6
Fuel fabrication	Reactor fuel
Power generation	Energy production
Reprocessing	Pu and U recovery from used fuel
Waste management	Storage, conditioning, and disposal of waste

$Z = 92$. It is possible to extract by milling and chemical leaching. Output is uranium ore concentrate (known as **yellowcake**, a U_3O_8 product). Under **conversion and enrichment stages**, we convert yellowcake to UF_6. Only ^{235}U is fissile (0.711% abundance). Generally, this can be enriched up to ~5 wt% in fuel by using gaseous diffusion or centrifuge of UF_6. Electromagnetic and laser techniques have been proposed and developed for enrichment process as well. In the **fuel fabrication stage**, enriched $^{235}UF_6$ is converted to UO_2. Then uranium oxide is then made into pellets (sintering and pressing). These pellets will then be loaded into fuel pins. These pins bundled will be formed into assemblies.

10.18 BACK-END STAGES

The important objective of the back end is to *treatment of used nuclear fuel*. First, the used nuclear **fuel storage** occurs in pools for cooling and shielding of a minimum of 2–3 years. For an interim storage, used nuclear fuel will be shipped in casks to dry cask storage area. The **reprocessing** stage is responsible in (1) recovery of fissile material to reuse for fabrication of new fuel and (2) separation of nuclear waste. Here, used fuel is composed of approximately: 96% U, 1% Pu, 3% FP + MA. Fissile U and Pu are valuable.

This can be done by recovered through dissolution using solvent extraction and reprocessing by dry method known as molten salt electrochemical methods. Details are outside of the scope of this textbook and should be consulted with the Nuclear Fuel Cycle course.

Waste management is another key important step. Here, used nuclear fuel must be suitably packaged for disposal. Wastes from reprocessing must be immobilized. Classify wastes depending on activity and heat (International Atomic Energy Agency (IAEA) classification but varies according to countries) to minimize volume and reduce hazard. Two forms of immobilization are cementation and vitrification. Lastly, geological disposal must be designed to complete the back-end stage to provide a high level of long-term isolation and containment. There will be no retrieval of waste afterward. Several countries are currently pursing geological disposal of a variety of waste types.

Important Terms (in the Order of Appearance)

Light water reactor	Fuel assembly/element	Generation II
Reactor pressure vessel	Control rods	Generation III
Coolant	Reactor trip/SCRAM	Generation IV
Thermal conversion efficiency	Decay heat	Nuclear fuel cycle
CANDU	Reactivity feedback	Open/Once-through fuel cycle
RBMK	Heavy water reactor	Closed fuel cycle
Molten salt reactor	Gas-cooled reactor	Mining and milling
Nuclear steam supply system	High-temperature gas-cooled reactor	Yellowcake
Balance of plant	Breeder reactors	Conversion and enrichment
Boiling water reactor	LMFBR	Fuel fabrication
Pressurized water reactor	GCFR	Fuel storage
Enrichment	MSBR	Reprocessing
Fuel rods/pins	Generation I	Waste management

BIBLIOGRAPHY

Bush, H.D., *Atomic and Nuclear Physics,* Prentice Hall, Englewood Cliffs, NJ, 1962.

Coombe, R.A., *An Introduction to Radioactivity for Engineers.* Macmillan Publishing Co., Inc., New York, 1968.

Duderstadt, J.J., and L.J. Hamilton, *Nuclear Reactor Analysis,* Wiley, New York, 1976.

El-Wakil, M.M, *Nuclear Power Engineering.* McGraw-Hill Book Company, Inc., New York, 1962.

Glasstone, S., and M.C. Edlund, *The Elements of Nuclear Reactor Theory*, D. Van Nostrand Company, New Jersey, 1952.

Glasstone, S., and A. Sesonske, *Nuclear Reactor Engineering*, 3rd ed., Van Nostrand Reinhold, New York, 1981

Glasstone, S., and R.H. Loveberg, *Controlled Thermonuclear Reactions,* D. Van Nostrand Company, Princeton, NJ, 1960.

Lamarsh, J.R., and Baratta, A.J., *Introduction to Nuclear Engineering*, 3rd Ed., Prentice Hall, NJ, 2001.

Lewis, E.E., *Nuclear Reactor Physics,* Elsevier, New York, 2008.

Mayo, R.M., *Nuclear Concepts for Engineers*, American Nuclear Society, La Grange Park, IL, 1998.

Shultis, J.K. and R.E. Faw, *Fundamentals of Nuclear Science and Engineering 2nd Ed.,* CRC Press, Taylor & Francis Group, Boca Raton, FL, 2008.

FURTHER EXERCISES

A. True or False: If the statement is false, give a counterexample or explain the correction. If the statement is true, explain why it is true.

    1. CANDU reactors use natural uranium as its fuel source. There have been no tests with enriched uranium fuel for CANDU designs.

2. Uranium metal fuel is mainly used for LWRs.
3. LWR has two types—PWR and BWR—both have control rods coming down from the top of the reactor.
4. Thermal conversion efficiency is the ratio of the electric power to thermal power.
5. GCR has higher thermal conversion efficiency than LWR.
6. In breeder reactors, it may begin operating with U-235 to give off radiation in order to initiate U-238 to breed Pu-239.
7. One of many advantages in using light water as a moderator is its cost—inexpensive.
8. One of many disadvantages in using light water as a moderator is its high boiling point.
9. CANDU reactors use heavy water as a moderator.
10. MSBR is the Gen III reactor concept.

B. Problems
1. A nuclear operator obtains the rate of heat rejected from a condenser to cooling water to be 2000 MW. This nuclear power plant has a thermal conversion efficiency of 34%.
a. Determine the electric and thermal power values for this nuclear power plant.
b. What is the flow rate (kg/s) of the condenser cooling water if the temperature rise of this water is 15°C? Data: Specific heat of water is ~4180 J/(kg°C).
2. Discuss the advantages and disadvantages for using (a) light water and (b) heavy water as a moderator in a nuclear reactor.
3. A reactor core using uranium rods is covered with a heavy water moderator. An engineer is proposing to replace it by light water moderator. By assuming that it is homogeneous in nature: (a) Would the thermal utilization increase or decrease? Why? (b) What would you expect the net effect on k_∞ to be? Why?

The next seven problems will be multiple-choice questions. There is only ONE correct answer.
Using the following partial nuclear fuel cycle diagram for Questions 4 and 5

$$\text{Mining/Milling} \xrightarrow{\text{X}} \text{Conversion} \xrightarrow{\text{Y}} \text{Enrichment}$$

4. The material X should be
a. U_3O_8
b. UO_2
c. UF_4
d. UF_6
e. None of the choices

5. The material Y should be
   a. U_3O_8
   b. UO_2
   c. UF_4
   d. UF_6
   e. None of the choices
6. It is possible to enrich uranium by _____.
   a. Gaseous diffusion
   b. Centrifugal
   c. Electromagnetic
   d. All choices
7. Which Gen II reactor type does not have enrichment cost?
   a. LMFBR
   b. CANDU
   c. PWR
   d. BWR
   e. AP1000
8. In a BWR, core reactivity is controlled by
   a. Coolant flow rate
   b. Soluble boron
   c. Control rods
   d. (a) and (c)
   e. (a) and (b)
   f. All of the choices
9. Boron worth is a technique used in
   a. All LWRs
   b. PWRs
   c. BWRs
   d. CANDU
   e. MSR
10. Conversion in nuclear fuel cycle implies that
    a. There is a transformation of natural uranium to uranium tetrafluoride.
    b. There is a transformation of U_3O_8 to UF_4.
    c. There is a transformation of U_3O_8 to UO_2.
    d. There is a transformation of U_3O_8 to UF_6.

11 Radiation Doses and Hazards

OBJECTIVES

After studying this chapter, the reader should be able to:

1. Define absorbed and equivalent dose and perform basic calculations in SI and traditional units.
2. Describe radiation damage mechanisms in materials and biological tissue for various particles.
3. Explain each of the three primary dose reduction principles.

11.1 PHYSIOLOGICAL EFFECTS

About 34 eV of energy is required on average to create an ion pair (ip) in soft tissue and air. For gases, this figure (W = 34 eV/ip) is reasonably independent of the type of ionizing radiation and its energy. Part of the deposited energy goes into molecular excitation and the formation of new chemicals. For example, water in cells can be converted into free radicals such as H, OH, H_2O_2, and HO_2. Direct damage can also occur in which radiation strikes certain molecules of the cells.

Example 11.1

Determine the charge deposition in tissue from a single 4 MeV alpha particle.

Solution

As each ion pair requires 34 eV, the alpha particle creates a large number of ion pairs before stopping.

$$N_{ip} = E_{\alpha}/W = 4 \times 10^6 \text{ eV}/(34 \text{ eV/ip}) = 1.2 \times 10^5 \text{ ion pairs.}$$

Each ion pair represents the release of one fundamental charge unit. The charge accumulation is therefore,

$$Q = eN_{ip} = (1.6 \times 10^{-19} \text{ C/ip})(1.2 \times 10^5 \text{ ip}) = \mathbf{1.9 \times 10^{-14}C.}$$

If a radioactive substance enters the body, a radiation exposure to organs and tissues will occur. However, the radioisotope will not deliver all of its energy to the body because of partial elimination, that is, the excretion of substances taken in by

DOI: 10.1201/9781003272588-11

inhalation and ingestion. If there are N atoms present, the radioactive decay rate is λN, and the biological elimination rate is $\lambda_B N$, then the total elimination rate of the radioisotope within the body is $dN/dt = -\lambda_E N = -\lambda N - \lambda_B N$, where the effective decay constant of elimination is $\lambda_E = \lambda + \lambda_B$. The fraction of the radioisotope that is removed by the body before decay is λ_B/λ_E, and the fraction decaying within the body is λ/λ_E. The corresponding relation between half-lives is then $1/t_E = 1/T_{1/2} + 1/t_B$, where each half-life is related to its decay constant via $t_k = \ln(2)/\lambda_k$.

Example 11.2

Calculate the effective half-life of ^{131}I that has an 8-day physical half-life and a 4-day biological half-life for the thyroid gland.

Solution

Thus, the effective half-life is

$$t_E = [(1/T_{1/2}) + (1/t_B)]^{-1} = [(1/8\,d) + (1/4\,d)]^{-1} = \mathbf{2.67\,d}.$$

11.2 NUCLEAR FISSION RADIATION SOURCES

Fission fragments, prompt neutrons, and γ radiation are emitted at the *time of fission*. Activation γ radiation results from (n, γ) reactions. This is present only when there is a neutron source. Delayed neutrons are emitted from certain fission fragments and spontaneous-fission neutrons from transmutation products. Only delayed neutrons exist after the fission process (i.e., the nuclear chain reaction) is terminated. Direct effects generally disappear within tens of minutes after the fission process is terminated because of short half-lives. Delayed α, β, and γ radiations with half-lives from fractions of a second to millions of years are emitted from fission fragments, activation products, and transmutation products. For example, ^{240}Pu is a long-lived spontaneous-fission source ($T_{1/2} = 6564$ y) in all irradiated reactor fuel that contains ^{238}U. Therefore, substantial concentrations of ^{240}Pu require continuous neutron shielding after the fission reaction is terminated.

11.3 ABSORBED DOSE

For a given volume of matter of mass (m), the energy (ΔE) "imparted" in some time interval is the sum of the energies (excluding rest-mass energies) of all charged and uncharged ionizing particles entering the volume minus the sum of the energies (excluding rest-mass energies) of all charged and uncharged ionizing particles leaving the volume, further corrected by subtracting the energy equivalent of any increase in rest-mass energy of the material in the volume. In other words, the energy imparted is that which is involved in the ionization and excitation of atoms and molecules within the volume and the associated chemical changes. This energy is eventually degraded almost entirely into thermal energy. Thus, **absorbed dose** is the quotient of the *mean* energy imparted ΔE to matter of mass Δm, in the limit as the mass approaches zero; that is,

$$D \equiv \lim_{\Delta m \to 0} \frac{\Delta E}{\Delta m}.$$

where D is the average energy absorbed from the radiation field per unit differential mass of the medium. Unit of absorbed dose is J/kg, typically referred to as a **gray** (Gy).

11.4 RADIATION EFFECTS

Older units are the **roentgen** and **rad**. 1 roentgen (R) equals the amount of radiation (X-ray or γ-ray) required to produce 1 esu (electrostatic unit) of charge from either part of an ion pair in $1\,cm^3$ of air at STP (standard temperature and pressure). The **rad** (radiation absorbed dose) equals the energy disposition per unit mass of material equal to 100 erg/g.

$$1 \text{ Gy (gray)} = 100 \text{ rad} = 1 \text{ J/kg}.$$

What we really want to calculate is not the amount of energy deposited but the biological effect on a human of that amount of energy being deposited for a particular the type of radiation. This is termed the RBE (**relative biological effectiveness**), and it depends on the effect studied, the dose, the dose rate, the physiological condition of the subject, etc.

The RBE is defined as: $\text{RBE} = \dfrac{\text{Dose of 250-keV x-rays producing given effect}}{\text{Dose of reference radiation for the same effect}}.$

Here, the **QF** (**quality factor**) is an *upper* limit of the RBE for a specific type of radiation. The **absorbed dose**, D, in rad or Gy, is related to **dose equivalent**, **H**, in rem or Sv, by

$$H = QF \times D.$$

The older unit of measurement of dose equivalent is the **rem**, which is an acronym for **Roentgen equivalent man**. The newer unit, the **sievert (Sv)**, is equivalent to 100 rem; that is, (1) Sv (sievert) = 100 rem, (2) H(rem) = D(rad) × QF, and (3) H(Sv) = D(Gy) × QF. The dose equivalent H is the absorbed dose D at a point in tissue weighted by a QF determined from the LET (linear energy transfer) of the radiation at that point. The LET is described in Chapter 8.

Similarly, the equivalent dose rate H′ and absorbed dose rate D′ are related by

$$H' = D' \times QF.$$

The QF essentially converts Gy to Sv or rad to rem (see Table 11.1). It should be noted that *equivalent doses and dose rates in Sv and rem are additive, independent of the specific radiation types involved.*

Important: *Values of quality factors will vary depending on the reference.*

TABLE 11.1

Quality Factor (QF)

Radiation	QF
X-rays, γ-rays, β⁻, β⁺	1
Neutrons <10 keV	5
10–100 keV	10
0.1–2 MeV	20
2–20 MeV	10
>20 MeV	5
Protons (>2 MeV) (ICRP)	5
Protons (>2 MeV) (NCRP)	2
Alpha particles and multiple-charged particles	20

Example 11.3

A beam of 1 MeV γ-rays having an intensity of 10^5 γ/cm²-s deposits in tissue 5×10^{-3} ergs/g-s. Calculate the absorbed dose and the dose equivalent rate.

Solution

By the definition of a rad, it takes a deposition of 100 erg/g-s to give 1 rad/s.
The absorbed dose rate D' is

$$D' = (5 \times 10^{-3} \text{ ergs/g-s})(3600 \text{ s/h})(1 \text{ rad}/100 \text{ erg/g}) = 0.18 \text{ rad/h} = \textbf{180 mrad/h}$$

or

$$D' = (0.18 \text{ rad/h})(1 \text{ Gy}/100 \text{ rad})(10^3 \text{ mGy/Gy}) = \textbf{1.8 mGy/h}.$$

Since for 1 MeV γ-rays, QF = 1, the dose equivalent rate is

$$H' = D' \times QF = D' \text{ (1 mrem/mrad)} = \textbf{180 mrem/h} \text{ or } \textbf{1.8 mSv/h}.$$

Example 11.4

Calculate the dose rate at a distance of 1 mile from a nuclear power plant if the dose rate at the plant boundary, ¼-mile radius, is 5 mrem/y.

Solution

Neglecting material attenuation in air, the inverse-square reduction factor is 1/16, according to

$$\frac{\varphi_2}{\varphi_1} = \frac{S/\left(4\pi r_2^2\right)}{S/\left(4\pi r_1^2\right)} = \frac{r_1^2}{r_2^2} = \frac{(0.25\text{mi})^2}{(1\text{mi})^2} = \frac{1}{16}.$$

Since the dose is directly proportional to the flux, the dose rate a mile from the plant is

$$D_2' = D_1'(r_1/r_2)^2 = (5 \text{ mrem/y})/16 = \mathbf{0.31 \text{ mrem/y}}.$$

Air attenuation would further reduce the dose to a negligible value.

The above calculation assumes that the measured dose rate is due entirely to radiation emissions from the nuclear power plant. **Background radiation** is the *naturally* occurring ionizing radiation in an environment independent of a specific source being measured, the latter typically referring to some manmade source of radiation such as a nuclear reactor. In the continental United States, the average natural background radiation level is about 300 mrem/y. If the 5 mrem/y in the above example is the regulatory limit for radiation emissions from a nuclear power plant at the plant boundary, it is reasonable to argue that the violations of the limit are difficult to determine with any certainty considering how they would be lost in a minimal variation to the natural background value.

Example 11.5

A small (point) source of radiation is located in the vacuum of space. At a distance of 20 m from the source a dose rate of 50 mGy/hr is measured, whereas at a distance of 50 m the measured dose rate is 35 mGy/hr. What is the background dose rate?

Solution

Let B be the background dose rate, D_1' be the dose rate at 20 m, and D_2' the *observed* dose rate at 50 m. The observed dose rate at a specific location is equal to the dose from the point source at that distance (weakened due to geometric attenuation) plus the background. If S_p is the point source value, that is, particles per second, $r_1 = 20$ m, and $r_2 = 50$ m, then

$$D_1' = S_p/(4\pi r_1^2) + B = 50 \text{ mGy/hr at } r_1,$$

$$D_2' = S_p/(4\pi r_2^2) + B = 35 \text{ mGy/hr at } r_2.$$

We have two equations with two unknowns, S_p and B, where we want to find the value for B in absorbed dose units of mGy/hr. We solve each equation for S_p and set the results equal to each other; that is, $S_p = 4\pi r_1^2(D_1' - B) = 4\pi r_2^2(D_2' - B)$. We now have a single equation with a single unknown, B. The 4π values cancel and we have

$$r_1^2(D_1' - B) = r_2^2(D_2' - B).$$

Substituting for the knowns,

$$(20 \text{ m})^2(50 \text{ mGy/hr} - B) = (50 \text{ m})^2(35 \text{ mGy/hr} - B)$$

$$50 \text{ mGy/hr} - B = 6.25(35 \text{ mGy/hr} - B) \rightarrow 50 \text{ mGy/hr} - B = 219 \text{ mGy/hr} - 6.25 B$$

$$5.25 B = 169 \text{ mGy/hr}.$$

Therefore, the background is **B = 32.2 mGy/hr.**

11.5 WHAT IS KERMA?

The **kerma** (stands for *k*inetic *e*nergy of *r*adiation absorbed per unit *ma*ss), in J/kg or Gy, is a deterministic absorbed dose quantity used only in connection with *indirectly ionizing* (uncharged) radiation such as neutrons and photons. Kerma is the sum of the initial kinetic energies per unit mass of all charged particles produced by the radiation.

Let E_{tr} be the sum of the initial kinetic energies of the primary ionizing particles (e.g., photoelectrons, Compton electrons, positron–electron pairs, nuclei scattered by fast neutrons) released by interaction of indirectly ionizing particles (X-rays, gamma rays, fast neutrons) in matter of mass m,

$$K \equiv \lim_{\Delta m \to 0} \frac{\Delta \langle E_{tr} \rangle}{\Delta m},$$

where K is (# of interactions per unit mass) × (energy imparted per interaction) and can be expressed mathematically by

$$K = [\Phi(\mu(E)/\rho)]\, E\, f(E),$$

where K is the kerma, Φ is the radiation fluence of energy E, which is $\Phi = \int_0^t \phi(t)\,dt$, f(E) is the fraction of the incident radiation particle's energy E transferred to secondary charged particles, and $\mu(E)/\rho$ = mass interaction coefficient for the detector material. It should be noted that for constant flux ϕ_0, $\Phi = \phi_0 t = Nvt$.

11.6 ABSORBED DOSE VERSUS KERMA

The *kerma* is a measure of all the energy transferred from the uncharged particle to primary ionizing particles per unit mass, whereas *absorbed dose* is a measure of the energy absorbed per unit mass. *Absorbed dose* and *kerma* are not necessarily equal since all the energy transferred to the primary ionizing particles in a given volume may **not** be absorbed in that volume (see Figure 11.1). For example, some of the energy may leave a given volume and be absorbed elsewhere as a result of Bremsstrahlung or annihilation radiation generated by the primary ionizing particles.

The absorbed dose equals the kerma under *charged particle equilibrium*, which is achieved in a small incremental volume about the point of interest if, for every charged particle leaving the volume, another of the same type with the same kinetic energy enters the volume traveling in the same direction. In a large medium, where electronic equilibrium exists and where there is insignificant energy loss by Bremsstrahlung, the *kerma* equals the *absorbed dose*.

Example 11.6[1]

A 10 MeV photo penetrates a 100 g mass and undergoes a single pair-production interaction, producing a positron and electron each of 4.5 MeV. Both

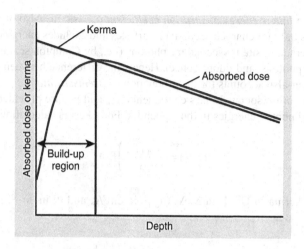

FIGURE 11.1 Relations between kerma and absorbed dose for photon radiation and fast neutrons. The absorbed dose initially increases as *electronic equilibrium* is approached and the ionization density increases due to the increasing number of secondary ions produced.

charged particles dissipate all their energy with the mass through ionization and Bremsstrahlung production. Three Bremsstrahlung photons of 1.6, 1.4, and 2 MeV each that are produced escape from the mass before they interact. The positron, after expending all its kinetic energy, interacts with an ambient electron within the mass, and they mutually annihilate one another to produce two photons of 0.51 MeV each, with both escape before they can interact within the mass. Calculate the kerma and the absorbed dose.

Solution

The total initial kinetic energy is entirely from the created positron–electron pair = 2 × (4.5 MeV) and K = (2 × 4.5 Me)(1.6 × 10⁻¹³ J/MeV)/(0.1 kg × 1 J/kg/Gy) = **1.44 × 10⁻¹¹ Gy.**

The absorbed energy is the initial kinetic energy minus that converted to Bremsstrahlung (1.6 MeV + 1.4 MeV + 2 MeV = 5.0 MeV) and into two photons of 2 × 0.51 MeV annihilation radiation, that is 1.02 MeV.

$$D = \frac{(10\,\text{MeV} - 5.0\,\text{MeV} - 1.02\,\text{MeV})\left(1.6 \times 10^{-13}\,\dfrac{\text{J}}{\text{MeV}}\right)}{\left(0.1\,\text{kg} \times 1\dfrac{\text{J/kg}}{\text{Gy}}\right)} = 6.37 \times 10^{-12}\,\text{Gy.}$$

11.7 PHOTON KERMA AND ABSORBED DOSE

Above we defined f, the fraction of the photon's energy E that is transferred to secondary charged particles.

$$K = [\Phi(\mu(E)/\rho)]\,E\,f(E)$$

We define the *linear energy absorption coefficient* μ_{en} as $\cong f\mu$. Also of interest is μ_{tr} that accounts only for charged secondary particles and excludes energy carried away from the interaction site by secondary photons (i.e., by Compton scattered photons, annihilation photons, and fluorescence). Hence, the difference between μ_{en} and μ_{tr} is that the former also accounts for the production of *Bremsstrahlung.*

Values of both associated mass coefficients, μ_{en} and μ_{tr}, are provided for various materials and photon energies in the Appendix. For a kerma calculation, then

$$K(E) = \left(\frac{\mu_{tr}(E)}{\rho} \right) E\Phi.$$

For units of kerma in Gy, E in MeV, (μ_{tr}/ρ) in cm^2/g, and Φ in cm^{-2}, this equation yields

$$K(E) = 1.602 \times 10^{-10} \left(\frac{\mu_{tr}(E)}{\rho} \right) E\Phi, Gy.$$

In the event of the production of significant Bremsstrahlung radiation by the secondary charged particles, we simply substitute μ_{en} for μ_{tr} in the previous equation.

11.8   KERMA FOR FAST NEUTRONS

The primary means of kinetic energy transfer by fast neutrons to a medium is by scattering interactions. Here, for isotropic elastic scattering in the center-of-mass system of a neutron with initial kinetic energy E, the average neutron energy loss, and therefore, the average energy imparted to the recoil nucleus with which the neutron interacts, is $f_s(E) = \dfrac{2A}{(A+1)^2} E$, where A is the atomic mass number of the scattering nucleus.

Assuming only elastic scattering to be important, the neutron kerma is therefore

$$K(E) = 1.602 \times 10^{-10} \left(\frac{f_s(E)\Sigma_s(E)}{\rho} \right) E\Phi,$$

where Σ_s is the macroscopic elastic scattering cross section in units of cm^{-1}. Here K is in units of Gy, *E* in MeV, and Φ in cm^{-2}.

11.9   RADIATION DAMAGE

The three main mechanisms of **radiation damage** are: (1) displacement of electrons and atoms—"knock-on" atoms, (2) large energy release in small volumes, and (3) production of impurities by neutron fission and activation. Alpha particles result in the buildup of He gas which, although chemically insert, can lead to over-pressurization problems. An example is the (n, α) reactions in^{10}B control-rod material in nuclear reactors.

Radiation damage increases in the following order of molecular formation:

1. Metallic bond
2. Ionic bond
3. Covalent bond (prevalent in biological tissue)
4. Van der Waals' bond.

Radiation damage is less at elevated temperatures where enhanced diffusion may provide repair mechanisms. Low melting points enhance "annealing" (i.e., the migration of dislocated atoms) and reduce effects. Damage is a function of both *dose* and *dose rate*.

11.10   BIOLOGICAL SPECIFIC EFFECTS

Cell damage is greatest in cells that have the highest degree of differentiation or rapid multiplication. The hierarchy of highest to lowest susceptibility is:

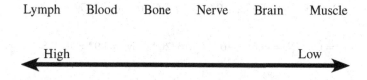

The fact that human senses cannot detect radiation may be of evolutionary significance in that from the aspect of geological time, typical levels of environmental radiation have never been hazardous.

11.11   RADIATION EFFECTS ON REACTOR MATERIALS

Key neutron interactions with respect to radiation effects on reactor materials are: fission fragments, fast neutrons, γ-rays, delayed radiations from fission, activation, and transmutation. The magnitude of damage is a function of neutron flux $\phi(t)$ history, i.e., neutron fluence Φ, which is $\Phi = \int \phi(t)dt$ with units of cm^{-2}.

The water moderator of LWRs is subject to dissociation in a radiation field where the H_2 and O_2 products enhance corrosion. Graphite reactor moderators swell (i.e., the volume increases while density decreases). This in turn leads to increased thermal resistance and internal energy. Finally, radiation effects on metal structures are: (1) hardening and embrittlement (e.g., the reactor vessel), (2) swelling, (3) change of metallurgical phase, (4) decreased corrosion resistance, and (5) change in mechanical properties.

11.12   MEASURING RADIATION EFFECTS

How much radiation is produced is measured by the *activity*; that is, decays per unit time with units of Curie (Ci) or Becquerel (Bq). How much of the radiation's energy is absorbed by tissue is quantified by the *dose* reported. Finally, how much biological damage does the radiation effect per energy absorbed is quantified by the **dose equivalent.**

Example 11.7

What is the dose equivalent rate in tissue, per picocurie, of an isotope emitting one 6-MeV alpha particle per decay? Given data: QF = 20 for a particles.

Solution

Assume that the a particle kinetic energy is deposited uniformly in the lungs with a mass of 1 kg.

$$D = \text{(energy deposited)/mass} = (6 \text{ MeV})(1.602 \times 10^{-13} \text{ J/MeV})/(1 \text{ kg}),$$

$$D = 9.61 \times 10^{-13} \text{ J/kg} = 9.61 \times 10^{-13} \text{ Gy/}\alpha \text{ (Note: 1 Gy = 1 J/kg)},$$

$$H = D \times QF = (9.61 \times 10^{-13} \text{ Gy/}\alpha) (20 \text{ Sv/Gy}) = 1.92 \times 10^{-11} \text{ Sv/}\alpha.$$

If the total activity is 1 pCi (pico = 10^{-12}), the dose equivalent rate is

$$(1.92 \times 10^{-11} \text{ Sv/}\alpha) (10 - 12 \text{ Ci})(3.7 \times 10^{10} \text{ a/s/Ci}) = 7.10 \times 10^{-13} \text{ Sv/s},$$

which is equivalent to $(7.10 \times 10^{-13} \text{ Sv/s})(100 \text{ rem/Sv})(3600 \text{ s/h}) = \mathbf{2.56 \times 10^{-7} \text{ rem/h}}.$

11.13   TOXICITY

The concept of **toxicity** takes into account the sensitivity of humans to radioisotopes after inhalation or ingestion. Historically, internal dose limits were specified in terms of maximum permissible concentration in air and water. The resulting dose commitment was considered as being separate from the annual limit for external exposure.

The more recent approach is to define for each radionuclide: (1) **ALI**—annual limit of intake and (2) **DAC**—derived air concentration, where DAC (i, k) is the *derived air concentration* (μCi/cm^3) of isotope i, in air or water (index k), and the DAC values can be found in 10 CFR 20 (Code of Federal Regulations). Based on these definitions, **toxicity** is the volume (typical units of m^3) of air or water with which the radioactive waste must be diluted in order for continuous breathing of that air or drinking of that water to result in a dose that is less than that corresponding to the dose (*100 mrem/y*) from continuous exposure to a concentration equal to a DAC. Table 11.2 provides various radiation sources. If we let A(i) be the activity of isotope i, μCi, then toxicity(i, k) is A(i)/DAC(i, k).

TABLE 11.2
Everyday Radiation Sources

Annual dose (mrem/y)	Source
0.003	If all electricity was from nuclear
0.3	Sleeping with another person
5	Allowed dose rate at the "fence line" of a nuclear power station
24	Bone dose from coal unit stack gases
525	Grand Central Station
335+[a]	Background for person living at 5000 ft. elevation and taking 5 transcontinental air flights per year
135–150[a]	Average U.S. east coast at sea level

[a] Radon in homes adds 100–1000 mrem/y.

Example 11.8

Compare the toxicities of 1 Ci of tritium (DAC value in air of 2.0×10^{-5} µCi/cm³) and 1 Ci of ^{137}Cs (DAC value in air of 2.0×10^{-8} µCi/cm³).

Solution

$$\text{toxicity}(^3\text{H}) = (1.0 \times 10^6 \text{ µCi})/(2.0 \times 10^{-5} \text{ µCi/cm}^3) = 5.0 \times 10^{10} \text{cm}^3 = 5.0 \times 10^4 \text{m}^3$$

$$\text{toxicity}(^{137}\text{Cs}) = (1.0 \times 10_6 \text{ µCi})/(2.0 \times 10^{-8} \text{ µCi/cm}^3) = 5.0 \times 10^{13} \text{cm}^3 = 5.0 \times 10^7 \text{m}^3$$

Thus, the ratio is toxicity(^{137}Cs)/ toxicity(^{3}H) = **1000**.

11.13 GENERAL RADIATION SAFETY

We will start with **ALARA**—*As Low as Reasonably Achievable*. The concept of ALARA assumes the linear no threshold (LNT) hypothesis of radiation risk, that is, a radiation dose of any magnitude can be detrimental. External radiation exposure can be mitigated by any combination of the following: (1) time—minimizing time of exposure, (2) distance—maximizing the distance from the source increases the geometric attenuation since exposure falls off as $1/r^2$, and (3) Shielding—maximizing the thickness and optimizing the material increase the material attenuation. **Irradiation** represents the exposure to penetrating radiation. External irradiation does not make a person radioactive. And **radioactive contamination** occurs when radioactive material is deposited any place where it is not desired. A contaminated person or thing (including the environment) will be irradiated until the source of the radiation is removed.

There are some general observations on radiation. First, the average natural background for continental U.S. is 3.0 mSv/y. This is a significant increase over earlier estimates due to inclusion of the effects of radon gas. Second, it is estimated that only one in ten million human cancers are caused by radiation, natural, or man-made. Third, the LNT ("Linear No Threshold") model has generally been used to establish exposure limits. The LNT model uses the concept of *collective dose*; for example, if 1000 rem can kill one person, then one rem to each of 1000 people can cause one fatality, as well as 1 mrem to each of one million people. Does this make any logical sense? Well, the LNT model extrapolates effects from high-dose experiments which are often known to have conservative biases. For example, mice are known to be more sensitive and differently sensitive to radiation than humans. Moreover, inbred mice and rats used for laboratory tests have vulnerable immune systems. Fourth, over 300 billion of the body's cells are struck by radiation from natural sources every day. Lastly, natural radiation causes 70 million DNA-damaging events in a person annually.

Hormesis is the concept that toxic materials are beneficial at low doses. Tolerable challenges to an organism stimulate the immune system and strengthen the organism. ("Any fight you win makes you stronger.") When shielded from external radiation and fed food that reduced their body's natural radiation, laboratory mice failed to function properly. When exposure to natural radiation was restored, they returned to normal functioning. Within the range of radon levels measured (up to 6 pCi/L), the number of lung cancer deaths *decreased* with increasing radon levels. This was opposite of the effect predicted by the LNT model. Figure 11.2 illustrates all these conceptual effect versus dose.

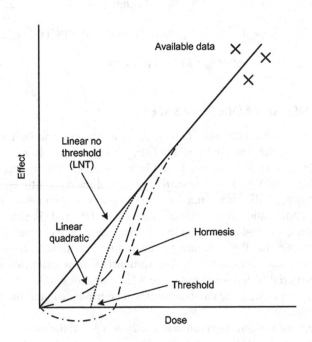

FIGURE 11.2 Differences between radiation dose and biological effect for various dose effect models.

11.14   RADIATION PROTECTION STANDARDS AND RISKS

Based on National Council of Radiation Protection and Measurements (NCRP) Statement No. 10, December 2004, "Recent Applications of the NCRP Public Dose Limit Recommendation for Ionizing Radiation," the principles of justification and "as low as reasonably achievable" ALARA have continued to be the overriding considerations of NCRP. The annual radiation dose limit recommendation for individual members of the public from all radiation sources other than natural background and the individual's medical care is for members of the public who are exposed continuously or frequently, 1 mSv/y (100 mrem/y) with an *infrequent* basis, 5 mSv/y (500 mrem/y). Here, *infrequent* is defined as a justified exposure that is not likely to occur often in an individual's lifetime, with each occurrence justified independently of any other. Typical natural background radiation is ~300 mrem/y, or 3 mSv/y.

Now, it is also possible to look at HPS (Health Physics Society) position statement: For doses below 100 mSv (10 rem), "...risks of health effects are either too small to be observed or are non-existent." Support for this position comes from results of investigation of survivors of the Hiroshima and Nagasaki atomic bombings. The results indicate that research on the bombing survivors found for the entire exposed population of survivors, an excess cancer mortality rate of about 0.67% and no detectable elevations in normal rates of illness and disease were found for survivor exposures below 100 mSv (10 rem). It is important to note that the HPS position dose is one hundred times that of the present radiation protection limit. Tables 11.3 and 11.4 provide maximum dose limits and estimate personal radiation dose from different sources to show various aspects of safety regulations.

"Is it *safe*?" and "Is it *dangerous*?" are not valid questions. Safety and risk are relative terms—that is, one must always ask, "Safe relative to what?" The following data and points of view are from *Creating the New World: Stories & Images from the Dawn of the Atomic Age* by Theodore Rockwell (see Table 11.5).

The numbers from Table 11.5 were obtained by dividing the total quantity of material produced by the estimated lethal dose.

The construction of a direct solar plant, that is, one producing electricity through the use of photovoltaic (PV) cells, requires 20 times more steel, 13 times more concrete, and twice as much construction labor as that of a nuclear plant of equal capacity. The PV cells which contain significant concentrations of toxic materials must eventually be disposed.

TABLE 11.3
Maximum Annual Occupational Dose Limits (Radiation Workers)

Body Parts	Maximum Dose Limits (mrem)
Whole body	5000
Extremities	50,000
Lens of the eye	15,000
Fetus (during the gestation period)	500
Individuals in the general public	100

Reference: NC State University, "Radiation Safety and ALARA"

TABLE 11.4

Annual dose Values at Various Locations with an Average Dose per Person from All Sources (~620 mrem/y)

Locations	Dose (mrem)
Cosmic radiation at sea level	28
Cosmic radiation at 5–6000 ft. above sea level	52
From ground in a state that borders the Gulf or Atlantic coasts	16
From ground in the Colorado Plateau	63
From ground anywhere else in the continental U.S.	30
Live in a stone, brick, or concrete building	7
Live within 50 miles of a nuclear power plant	0.01
Live within 50 miles of a coal-fired power plant	0.03
Internal radiation from food (C-14 and K-40)	40
Internal radiation from water (Rn dissolved in water)	40
Internal radiation from air (radon)	228
Smoke ½ pack of cigarettes per day	18
Chest X-ray	10
Mammography X-ray	42
Barium enema	800
Head Computerized Tomography (CT)	200
Chest CT	700
Whole-body CT	1000
Cardiac CT	2000

Reference: ANS "Estimate your personal annual radiation dose," 2011 (https://www.ans.org/nuclear/dosechart/)

TABLE 11.5

Production of Toxic Materials—Extreme Case Mortality Rates That Could Potentially Result from the Annual U.S. Production of Five Chemicals, All with *Infinite* Half-Lives

Chemical	Number of Human Deaths
Barium	100 billion
Hydrogen cyanide	6 trillion
Ammonia	6 trillion
Phosgene	20 trillion
Chlorine gas	400 trillion

Important Terms (in the Order of Appearance)

Absorbed dose	Rem	DAC
Gray (Gy)	Sievert	ALARA
Roentgen (R)	Background radiation	Irradiation
Rad	Kerma	Radioactive contamination
RBE	Radiation damage	Hormesis
Quality factor (QF)	Toxicity	
Dose equivalent	ALI	

BIBLIOGRAPHY

Cember, H. and T.E. Johnson, *Introduction to Heath Physics,* 4th ed. McGraw-Hill Education, 2008.

Foster, A.R. and R.L. Wright Jr., *Basic Nuclear Engineering 4th Ed.*, Allyn & Bacon, Newton, MA 1983.

Glasstone, S., and A. Sesonske, *Nuclear Reactor Engineering*, 3rd ed., Van Nostrand Reinhold, New York, 1981.

Lamarsh, J.R., and Baratta, A.J., *Introduction to Nuclear Engineering*, 3rd Ed., Prentice Hall, NJ, 2001.

Radiation Protection, ICRP Publication 26, Annals of the ICRP 1, No.3 (January 1977).

Radiological Health Handbook, U.S Department of Health, Education and Welfare, 1962.

Shultis, J.K. and R.E. Faw, *Fundamentals of Nuclear Science and Engineering 2nd Ed.,* CRC Press, Taylor & Francis Group, Boca Raton, FL, 2008.

Shultis, J.K. and R.E. Faw, *Radiation Shielding,* American Nuclear Society, La Grange Park, IL, 2000.

The Effects on Populations of Exposure to Low levels of Ionizing Radiation, National Academy of Sciences, July 1980.

The Effects on Populations of Exposure to Low Levels of Ionizing Radiation, NAS/NRC Report, November 1972.

FURTHER EXERCISES

A. True or False: If the statement is false, give a counterexample or explain the correction. If the statement is true, explain why it is true.

1. The exposure expressed in roentgens does not depend on time.
2. 1 rad is equivalent to 100 ergs/g or 0.01 J/kg for any material, any radiation.
3. 1 Gray = 1000 rad for any material and any radiation.
4. Gray and rem are often used as the units of dose for biological systems.
5. Quality factor is the upper limit of the relative biological effectiveness.
6. Quality factor for an alpha particle is smaller than that for a beta.
7. Cosmic radiation at the top of Teton, WY is higher than that at Richmond, VA.
8. Fetus is more sensitive to radiation than bone.

9. Three fundamental rules of radiation protection are time, distance, and shielding.
10. Kerma dose is often less than the absorbed dose.

B. Problems[2]

1. A worker in a nuclear laboratory receives a whole-body exposure for 5 minutes by a thermal neutron beam at a rate 20 mrad/h. What absorbed dose (in mrad) and dose equivalent (in mrem) does he receive? What fraction of the yearly dose limit of 5000 mrem/y for an individual is this?

2. A plant worker accidentally breathes some stored gaseous tritium, a beta emitter with maximum particle energy 0.0186 MeV. The energy absorbed by the lungs, of total mass 1 kg, is 4×10^{-3} J. How many millirem dose equivalent was received? How many millisieverts?

3. The nuclear reactions resulting from thermal neutron absorption in boron and cadmium are

$$^{10}B_5 + {}^1n_0 \rightarrow {}^7Li_3 + {}^4He_2$$

$$^{113}Cd_{48} + {}^1n_0 \rightarrow {}^{114}Cd_{48} + \gamma[5 \text{ MeV}].$$

Which of these two materials would be more effective radiation shield? Explain.

4. The radiation absorbed dose rate, D', from a source of X-rays varies with time according to the function $D_0' \sin(\pi t/T)$, where t is the elapsed time in seconds over which the subject is exposed, $D_0' = 5.2$ mGy/s, and T is the total time of exposure which is 540 seconds. What is (a) the total absorbed dose received in mGy and (b) the total equivalent dose received in Sv?

5. An astronaut in the vacuum of space is performing maintenance near a small reactor in orbit. Assume you can treat the reactor as a point source.

    a. If the astronaut's absorbed dose rate at a distance of 2 m from the reactor is 25 mGy/hr from gamma rays and 4 mGy/hr from 1 MeV fast neutrons, what is the astronaut's equivalent dose rate in mSv/hr at a distance of 10 m from the reactor assuming the only source of radiation is the reactor (i.e., ignore background radiation)?

    b. What is the astronaut's dose equivalent in mSv and in *mrem* after working for 30 minutes at a distance of 10 m from the reactor?

6. You have the three radioactive cookies shown. Assume all radioisotopes in the cookies are "whole-body-seekers" and have the same radiological and biological half-life.

a. Which cookie is safest to eat and <u>why</u>?

b. Which cookie is the least safe to eat and <u>why</u>?

7. Rank the following items from high to low sensitivity to radiation: fetus, bone, muscles, intestines, skin, and thyroid.

8. A person working in a dusty high radiation environment is exposed to 20 rad/h of gamma radiation and 15 rad/h of fast neutrons (2–20 MeV range) and 6 rad/h of thermal neutrons. How long, in second, can this person work within this environment without exceeding the recommended NCRP guideline of 3 rems in any 13-week periods?

9. What level of activity of tritium should be in order to have an equivalent toxicity to 1 Ci of ^{137}Cs? Data: tritium—DAC value in air of 2.0×10^{-5} µCi/cm^3 and ^{137}Cs—DAC value in air of 2.0×10^{-8} µCi/cm^3.

NOTES

1 This example is adapted from Example 6.8, Cember and Johnson, *Introduction to Health Physics*, 4th ed.

2 Assume QF values from Table 11.1 in this chapter.

Appendix I
Conversion Factors and Basic Atomic Data Sets

A   CONVERSION FACTORS

Length
1 km = 0.6215 mi
1 mi = 1.609 km
1 m = 1.0936 yd = 3.281 ft = 39.37 in
1 in = 2.54 cm
1 ft = 12 in = 30.48 cm
1 yd = 3 ft
1 light year = 9.461 × 10^{15} m

Speed
1 km/h = 0.2778 m/s = 0.6215 mi/h
1 mi/h = 1.467 ft/s = 0.4470 m/s

Angle and Angular Speed
π rad = 180°
1 rad = 57.30°
1 rev/min = 0.1047 rad/s

Force
1 N = 0.2248 lb = 10^5 dyn
1 lb = 4.4482 N

Pressure
1 Pa = 1 N/m²
1 atm = 101.325 kPa = 1.01325 bars
1 atm = 14.7 lb/in² = 760 mmHg
1 torr = 133.32 Pa
1 bar = 100 kPa

Thermal conductivity
1 W/m-K = 6.938 Btu-in/(h-ft² – °F)

Area
1 km² = 0.3861 mi² = 247.1 acres

Volume
1 gal = 3.786 L = 4 qt = 8 pt = 128 oz

Time
1 h = 60 min
1 d = 24 h = 1440 min
1 y = 365.24 d

Mass
1 kg = 1000 g = 2.20462 lbm
1 tonne = 1000 kg
1 u = 1.6606 × 10^{-27} kg = 931.50 MeV/c²
1 slug = 14.59 kg

Density
1 g/cm³ = 1000 kg/m³ = 1 kg/L

Power
1 horsepower = 550 ft-lb/s = 745.7 W
1 Btu/min = 17.58 W
1 W = 1.341 × 10^{-3} horsepower

Energy
1 kW-h = 3.6 MJ
1 cal = 4.1840 J
1 ft-lb = 1.356 J = 1.286 × 10^{-3} Btu
1 Btu = 252 cal = 1054.35 J
1 eV = 1.602 × 10^{-19} J

Cross section
1 barn = 10^{-24} cm²

B PERCENT ISOTOPIC ABUNDANCES OF ELEMENTS AND NUCLEAR DATA FOR VARIOUS ELEMENTS AND ISOTOPES

Nuclides marked with an asterisk (*) in the abundance column indicate that it is not present in nature. The isotopic mass data is from G. Audi, A. H. Wapstra Nucl. Phys A. 1993, 565, 1–65 and G. Audi, A. H. Wapstra Nucl. Phys A. 1995, 595, 409–480. The percent natural abundance data are from the 1997 report of the IUPAC Subcommittee for Isotopic Abundance Measurements by K.J.R. Rosman, P.D.P. Taylor Pure Appl. Chem. 1999, 71, 1593–1607. Half-life and neutron cross section data sets are obtained from A.R. Foster and R.L. Wright, "Basic Nuclear Engineering," 4th Edition, Allyn & Bacon, pp. 577–582. It should be noted that for the symbol "nX," this represents ^nX for the correct notation discussed in the textbook; and that would represent an isotope. For a plain X, that means it represents an element. Note: (x, y) is the displacement caused by nuclear bombardment reactions and sf represents spontaneous fission.

| | | | | | | Neutron Cross Section (Barns) | | |
| | | | | | | | Scattering | |
Z	Name	Symbol	Mass of Atom (u)	% Abundance	Half-Life	Absorption at 0.025 eV	Thermal	Epithermal
1	Hydrogen	H	1.00797			0.332	38	20.4
		1H	1.007825	99.9885		0.332	38	20.4
	Deuterium	2H (D)	2.014102	0.0115		0.00046	7	3.4
	Tritium	3H (T)	3.016049	*	12.6 y			
2	Helium	He	4.002603			0.007	0.8	0.83
		3He	3.016029	0.000137		5500	0.8	
		4He	4.002603	99.999863		0	0.8	
3	Lithium	Li	6.939			70	1.4	0.9
		6Li	6.015122	7.59		945(n,α)		
		7Li	7.016004	92.41		0.033		
4	Beryllium	9Be	9.012182	100		0.010	7	6.11
5	Boron	B	10.811			755	4	3.7
		10B	10.012937	19.9		3813(n,α)		
		11B	11.009305	80.1		$< 5 \times 10^{-2}$		
6	Carbon	C	12.01115			0.0034	4.8	4.66
		12C	12.000000	98.93				
		13C	13.003355	1.07		0.0005		
		14C	14.003242	*	5715 y			
7	Nitrogen	N	14.0067			1.88	10	9.9

(Continued)

Z	Name	Symbol	Mass of Atom (u)	% Abundance	Half-Life	Absorption at 0.025 eV	Scatter ing Thermal	Scatter ing Epithermal
							Neutron Cross Section (Barns)	
		14N	14.003074	99.632				
		15N	15.000109	0.368				
8	Oxygen	O	15.9994			0.00019	4.2	3.75
		16O	15.994915	99.757				
		17O	16.999132	0.038				
		18O	17.999160	0.205				
9	Fluorine	19F	18.998403	100		0.010	3.9	3.6
10	Neon	20Ne	19.992440	90.48				
		21Ne	20.993847	0.27				
		22Ne	21.991386	9.25				
11	Sodium	23Na	22.989770	100		0.53	4.0	3.1
12	Magnesium	Mg	24.312			0.063	3.6	3.4
		24Mg	23.985042	78.99		0.034		
		25Mg	24.985837	10.00		0.280		
		26Mg	25.982593	11.01		0.060		
13	Aluminum	27Al	26.981538	100		0.230	1.4	1.4
14	Silicon	Si	28.086			0.16	1.7	2.2
		28Si	27.976927	92.2297		0.080		
		29Si	28.976495	4.6832		0.280		
		30Si	29.973770	3.0872		0.40		
15	Phosphorus	31P	30.973762	100		0.20	5	3.4
16	Sulfur	S	32.065			0.52		
		32S	31.972071	94.93				
		33S	32.971458	0.76				
		34S	33.967867	4.29				
		36S	35.967081	0.02				
17	Chlorine	Cl	35.453			33.5		
		35Cl	34.968853	75.78				
		37Cl	36.965903	24.22		56		
18	Argon	Ar	39.948			0.66		
		36Ar	35.967546	0.3365				
		37Ar	36.96678	*	35.1 d			
		38Ar	37.962732	0.0632				
		40Ar	39.962383	99.6003				

(Continued)

Z	Name	Symbol	Mass of Atom (u)	% Abundance	Half-Life	Absorption at 0.025 eV	Scattering Thermal	Scattering Epithermal
							Neutron Cross Section (Barns)	
19	Potassium	K	39.102			2.07	2.5	2.1
		39K	38.963707	93.2581				
		40K	39.963999	0.0117				
		41K	40.961826	6.7302				
20	Calcium	Ca	40.078			0.43		
		40Ca	39.962591	96.941				
		42Ca	41.958618	0.647				
		43Ca	42.958767	0.135				
		44Ca	43.955481	2.086				
		46Ca	45.953693	0.004				
		48Ca	47.952534	0.187	5×10^{19} y			
21	Scandium	45Sc	44.955910	100		27.2		
22	Titanium	Ti	47.90			6.1	4.0	4.2
		46Ti	45.952629	8.25				
		47Ti	46.951764	7.44				
		48Ti	47.947947	73.72				
		49Ti	48.947871	5.41				
		50Ti	49.944792	5.18				
23	Vanadium	V	50.942			5.0	5.0	
		50V	49.947163	0.250	1.4×10^{17} y			
		51V	50.943964	99.750				
24	Chromium	Cr	51.996			3.1	3.0	3.9
		50Cr	49.946050	4.345				
		52Cr	51.940512	.83.789				
		53Cr	52.940654	9.501		1.82		
		54Cr	53.938885	2.365				
25	Manganese	55Mn	54.938050	100		13.2	2.3	1.9
26	Iron	Fe	55.847			2.62	11	11.4
		54Fe	53.939615	5.845		2.3		
		56Fe	55.934942	91.754		2.7		
		57Fe	56.935399	2.119		2.5		
		58Fe	57.933280	0.282		1.2		
27	Cobalt	59Co	58.933200	100		37.2	7	5.8
28	Nickel	Ni	58.71			4.8	17.5	17.4
		58Ni	57.935348	68.0769				
		59Ni	58.934345	*	7.6×10^4 y			

(*Continued*)

Z	Name	Symbol	Mass of Atom (u)	% Abundance	Half-Life	Absorption at 0.025 eV	Scattering Thermal	Scattering Epithermal
		60Ni	59.930791	26.2231				
		61Ni	60.931060	1.1399				
		62Ni	61.928349	3.6345				
		64Ni	63.927970	0.9256				
29	Copper	Cu	63.54			3.77	7.2	7.7
		63Cu	62.929601	69.17		4.6		
		64Cu	63.929763	*	12.70 h			
		65Cu	64.927794	30.83		2.2		
30	Zinc	Zn	65.409			1.1		
		64Zn	63.929147	48.63				
		66Zn	65.926037	27.90				
		67Zn	66.927131	4.10				
		68Zn	67.924848	18.75				
		70Zn	69.925325	0.62				
31	Gallium	Ga	69.723			2.9		
		69Ga	68.925581	60.108				
		71Ga	70.924705	39.892				
32	Germanium	Ge	72.64			2.2		
		70Ge	69.924250	20.84				
		72Ge	71.922076	27.54				
		73Ge	72.923459	7.73				
		74Ge	73.921178	36.28				
		76Ge	75.921403	7.61	1.6×10^{21} y			
33	Arsenic	75As	74.921596	100		4.2		
34	Selenium	Se	78.96		9×10^{19} y	11.7		
		74Se	73.922477	0.89				
		76Se	75.919214	9.37				
		77Se	76.919915	7.63				
		78Se	77.917310	23.77				
		80Se	79.916522	49.61				
		82Se	81.916700	8.73	9×10^{19} y			
35	Bromine	Br	79.904			6.8		
		79Br	78.918338	50.69		2.45 and 8.4	6	6
		81Br	80.916291	49.31				
36	Krypton	Kr	83.798			24		
		78Kr	77.920386	0.35				

(*Continued*)

Z	Name	Symbol	Mass of Atom (u)	% Abundance	Half-Life	Absorption at 0.025 eV	Scattering Thermal	Epithermal
		80Kr	79.916378	2.28				
		82Kr	81.913485	11.58				
		83Kr	82.914136	11.49				
		84Kr	83.911507	57.00				
		86Kr	85.910610	17.30				
37	Rubidium	Rb	85.4678			0.39		
		85Rb	84.911789	72.17				
		87Rb	86.909183	27.83	4.88×10^{10} y	0.12		
38	Strontium	Sr	87.62			1.2		
		84Sr	83.913425	0.56				
		86Sr	85.909262	9.86				
		87Sr	86.908879	7.00				
		88Sr	87.905614	82.58				
		90Sr	89.9077308	*	28.78 y			
39	Yttrium	89Y	88.905848	100		1.28		
40	Zirconium	Zr	91.22			0.180	8	6.2
		90Zr	89.904704	51.45				
		91Zr	90.905645	11.22				
		92Zr	91.905040	17.15				
		94Zr	93.906316	17.38				
		96Zr	95.908276	2.80	2×10^{19} y			
41	Niobium	93Nb	92.906378	100		1.15	5	6.5
42	Molybdenum	Mo	95.94			2.7	7	6
		92Mo	91.906810	14.84				
		94Mo	93.905088	9.25				
		95Mo	94.905841	15.92				
		96Mo	95.904679	16.68				
		97Mo	96.906021	9.55				
		98Mo	97.905408	24.13				
		100Mo	99.907477	9.63	8×10^{18} y			
43	Technetium	98Tc	97.907216	*	4.2×10^{6} y			
44	Ruthenium	Ru	101.07				2.6	
		96Ru	95.907598	5.54				
		98Ru	97.905287	1.87				

(*Continued*)

| | | | | | | | Neutron Cross Section (Barns) | |
| | | | | | | | Scatter ing | |
Z	Name	Symbol	Mass of Atom (u)	% Abundance	Half-Life	Absorption at 0.025 eV	Thermal	Epithermal
		99Ru	98.905939	12.76				
		100Ru	99.904220	12.60				
		101Ru	100.905582	17.06				
		102Ru	101.904350	31.55				
		104Ru	103.905430	18.62				
45	Rhodium	103Rh	102.905504	100			145	
46	Palladium	Pd	106.42			7		
		102Pd	101.905608	1.02				
		104Pd	103.904035	11.14				
		105Pd	104.905084	22.33				
		106Pd	105.903483	27.33				
		108Pd	107.903894	26.46				
		110Pd	109.905152	11.72				
47	Silver	Ag	107.870			63	6	6.4
		107Ag	106.905093	51.839		31		
		109Ag	108.904756	48.161		87		
48	Cadmium	Cd	112.40			2520	7	
		106Cd	105.906458	1.25				
		108Cd	107.904183	0.89				
		110Cd	109.903006	12.49				
		111Cd	110.904182	12.80				
		112Cd	111.902757	24.13				
		113Cd	112.904401	12.22	7.7×10^{15} y			
		114Cd	113.903358	28.73				
		116Cd	115.904755	7.49	3×10^{19} y			
49	Indium	In	114.82			196	2.2	
		113In	112.904061	4.29				
		115In	114.903878	95.71	4.4×10^{14} y	50 and 150		
50	Tin	Sn	118.710			0.61		
		112Sn	111.904821	0.97				
		114Sn	113.902782	0.66				
		115Sn	114.903346	0.34				
		116Sn	115.901744	14.54				
		117Sn	116.902954	7.68				
		118Sn	117.901606	24.22				

(*Continued*)

Z	Name	Symbol	Mass of Atom (u)	% Abundance	Half-Life	Neutron Cross Section (Barns) Absorption at 0.025 eV	Scatter ing Thermal Epithermal
		119Sn	118.903309	8.59			
		120Sn	119.902197	32.58			
		122Sn	121.903440	4.63			
		124Sn	123.905275	5.79			
51	Antimony	Sb	121.760			5.1	
		121Sb	120.903818	57.21			
		123Sb	122.904216	42.79			
52	Tellurium	Te	127.60			4.2	
		120Te	119.904020	0.09			
		122Te	121.903047	2.55			
		123Te	122.904273	0.89			
		124Te	123.902819	4.74			
		125Te	124.904425	7.07			
		126Te	125.903306	18.84			
		128Te	127.904461	31.74			
		130Te	129.906223	34.08	2.4×10^{21} y		
53	Iodine	127I	126.904468	100		6.2	
54	Xenon	Xe	131.30			25	4.3
		124Xe	123.905896	0.09			
		126Xe	125.904269	0.09			
		128Xe	127.903530	1.92			
		129Xe	128.904779	26.44			
		130Xe	129.903508	4.08			
		131Xe	130.905082	21.18			
		132Xe	131.904154	26.89			
		134Xe	133.905395	10.44			
		135Xe	–		9.2 h	2.72×10^6	
		136Xe	135.907220	8.87		0.15	
55	Cesium	Cs	132.90545			29	
		133Cs	132.905447	100			
		137Cs	136.907084	*	30.07 y		
56	Barium	Ba	137.327			1.3	
		130Ba	129.906310	0.106			
		132Ba	131.905056	0.101			
		134Ba	133.904503	2.417			

(*Continued*)

Z	Name	Symbol	Mass of Atom (u)	% Abundance	Half-Life	Neutron Cross Section (Barns) Absorption at 0.025 eV	Scatter ing Thermal	Epithermal
		135Ba	134.905683	6.592				
		136Ba	135.904570	7.854				
		137Ba	136.905821	11.232				
		138Ba	137.905241	71.698				
57	Lanthanum	La	138.9055			9.0		
		138La	137.907107	0.090				
		139La	138.906348	99.910				
58	Cerium	Ce	140.116			0.63		
		136Ce	135.907144	0.185				
		138Ce	137.905986	0.251				
		140Ce	139.905434	88.450				
		142Ce	141.909240	11.114				
59	Praseodymium	Pr	140.907648			11.4		
		141Pr	140.907648	100				
60	Neodymium	Nd	144.24			50		
		142Nd	141.907719	27.2				
		143Nd	142.909810	12.2				
		144Nd	143.910083	23.8	2.38×10^{15} y			
		145Nd	144.912569	8.3				
		146Nd	145.913112	17.2				
		148Nd	147.916889	5.7				
		150Nd	149.920887	5.6	7×10^{18} y			
61	Promethium	145Pm	144.912744	*	17.7 y			
62	Samarium	Sm	150.35			5,600		
		144Sm	143.911995	3.07				
		147Sm	146.914893	14.99	1.06×10^{11} y			
		148Sm	147.914818	11.24	7×10^{15} y			
		149Sm	148.917180	13.82		40,800		
		150Sm	149.917271	7.38				
		152Sm	151.919728	26.75				
		154Sm	153.922205	22.75				
63	Europium	Eu	151.964			4,600		
		151Eu	150.919846	47.81				
		153Eu	152.921226	52.19				

(*Continued*)

Z	Name	Symbol	Mass of Atom (u)	% Abundance	Half-Life	Neutron Cross Section (Barns) Absorption at 0.025 eV	Scattering Thermal Epithermal
64	Gadolinium	Gd	157.25			49,000	
		152Gd	151.919788	0.20	$1.1, \times 10^{14}$ y		
		154Gd	153.920862	2.18			
		155Gd	154.922619	14.80		61,000	
		156Gd	155.922120	20.47			
		157Gd	156.923957	15.65		240,000	
		158Gd	157.924101	24.84			
		160Gd	159.927051	21.86			
65	Terbium	Tb	158.925343			23.2	
		159Tb	158.925343	100		23.2	
66	Dysprosium	Dy	162.500			950	
		156Dy	155.924278	0.06			
		158Dy	157.924405	0.10			
		160Dy	159.925194	2.34			
		161Dy	160.926930	18.91			
		162Dy	161.926795	25.51			
		163Dy	162.928728	24.90			
		164Dy	163.929171	28.18			
67	Holmium	165Ho	164.930319	100		61	
68	Erbium	Er	167.259			160	
		162Er	161.928775	0.14			
		164Er	163.929197	1.61			
		166Er	165.930290	33.61			
		167Er	166.932045	22.93			
		168Er	167.932368	26.78			
		170Er	169.935460	14.93			
69	Thulium	169Tm	168.934211	100		105	
70	Ytterbium	Yb	173.04			50	
		168Yb	167.933894	0.13			
		170Yb	169.934759	3.04			
		171Yb	170.936322	14.28			
		172Yb	171.936378	21.83			
		173Yb	172.938207	16.13			
		174Yb	173.938858	31.83			

(Continued)

Z	Name	Symbol	Mass of Atom (u)	% Abundance	Half-Life	Absorption at 0.025 eV	Neutron Cross Section (Barns) Scattering Thermal	Epithermal
		176Yb	175.942568	12.76				
71	Lutetium	Lu	174.967			84		
		175Lu	174.940768	97.41				
		176Lu	175.942682	2.59				
72	Hafnium	Hf	178.49			104		
		174Hf	173.940040	0.16	2×10^{15} y			
		176Hf	175.941402	5.26				
		177Hf	176.943220	18.60				
		178Hf	177.943698	27.28				
		179Hf	178.945815	13.62				
		180Hf	179.946549	35.08				
73	Tantalum	Ta	180.9479			21		
		180Ta	179.947466	0.012				
		181Ta	180.947996	99.988		21	5	6
74	Tungsten	W	183.84			18.2		
		180W	179.946706	0.12				
		182W	181.948206	26.50				
		183W	182.950224	14.31				
		184W	183.950933	30.64				
		186W	185.954362	28.43				
75	Rhenium	Re	186.207			89		
		185Re	184.952956	37.40				
		187Re	186.955751	62.60	4.12×10^{10} y			
76	Osmium	Os	190.23			15		
		184Os	183.952491	0.02				
		186Os	185.953838	1.59	2×10^{15} y			
		187Os	186.955748	1.96				
		188Os	187.955836	13.24				
		189Os	188.958145	16.15				
		190Os	189.958445	26.26				
		192Os	191.961479	40.78				
77	Iridium	Ir	192.217			420		
		191Ir	190.960591	37.3				
		193Ir	192.962924	62.7				

(Continued)

Z	Name	Symbol	Mass of Atom (u)	% Abundance	Half-Life	Absorption at 0.025 eV	Neutron Cross Section (Barns) Scatter ing Thermal	Epithermal
78	Platinum	Pt	195.078			10		
		190Pt	189.959930	0.014	6.5×10^{11} y			
		192Pt	191.961035	0.782				
		194Pt	193.962664	32.967				
		195Pt	194.964774	33.832				
		196Pt	195.964935	25.242				
		198Pt	197.967876	7.163				
79	Gold	197Au	196.966552	100		98.8	9.3	
80	Mercury	Hg	200.59			374		
		196Hg	195.965815	0.15				
		198Hg	197.966752	9.97				
		199Hg	198.968262	16.87				
		200Hg	199.968309	23.10				
		201Hg	200.970285	13.18				
		202Hg	201.970626	29.86				
		204Hg	203.973476	6.87				
81	Thallium	Tl	204.3833			3.4		
		203Tl	202.972329	29.524				
		205Tl	204.974412	70.476				
82	Lead	Pb	207.19			0.170	11	11.3
		204Pb	203.973029	1.4	1.4×10^{17} y	0.8		
		206Pb	205.974449	24.1				
		207Pb	206.975881	22.1				
		208Pb	207.976636	52.4				
		209Pb	208.9810	*	3.3 h			
83	Bismuth	209Bi	208.980383	100		0.019 & 0.015	9	9.28
84	Polonium	209Po	208.982416	*				
		210Po	209.9829	*	138 d			
		213Po	212.9928		4×10^{-6} s			
85	Astatine	210At	209.987131	*				
86	Radon	222Rn	222.017570	*				
87	Francium	223Fr	223.019731	*				
88	Radium	226Ra	226.025403	*	1620 y	20		
89	Actinium	227Ac	227.027747	*				
90	Thorium	Th	232.0381			7.56		

(Continued)

Z	Name	Symbol	Mass of Atom (u)	% Abundance	Half-Life	Absorption at 0.025 eV	Neutron Cross Section (Barns) Scattering		
							Thermal	Epithermal	
		230Th	230.0331	*	7.6×10^4 y				
		232Th	232.038050	100	1.45×10^{10} y	7.56	12.5	12.5	
		233Th	233.04158	*	22.1 min				
91	Protactinium	Pa	231.035879			200			
		231Pa	231.035879	100	3.28×10^4 y				
		233Pa	233.04024	*	27.4 d	43			
92	Uranium	U	238.0289			7.68	8.3		
		233U	233.0395	*	1.59×10^5 y	530(n,f)	8.2		
						46(n,γ)			
		234U	234.040946	0.0055	2.5×10^5 y	105			
		235U	235.043923	0.7200	7.04×10^8 y	582(n,f)	10		
						112(n,γ)			
		236U	236.0457	*	2.34×10^7 y	7			
		237U	237.048728	*	6.75 d				
		238U	238.050783	99.2745	4.47×10^9 y	2.71	8.3		
		239U	239.054292	*	23.5 min	14(n,f)			
93	Neptunium	237Np	237.048167	*	2.14×10^6 y				
		238Np	238.050944	*	2.117 d				
		239Np	239.052937	*	2.355 d				
94	Plutonium	238Pu	238.04958	*	87.8 y				
		239Pu	239.0522	*	24360 y	746(n,f)			
						280(n,γ)			
		240Pu	240.0540	*	6760 y	<0.1(n,f)			
						295(n,γ)			
		241Pu	241.056849	*	14.4 y	1025 (n,f)			
						375(n,γ)			
		242Pu	242.0587	*	3.79×10^5 y	<0.2(n,f)			
						30(n,γ)			

(Continued)

Z	Name	Symbol	Mass of Atom (u)	% Abundance	Half-Life	Neutron Cross Section (Barns)		
						Absorption at 0.025 eV	Scatter ing Thermal	Epithermal
		244Pu	244.064198	*	8.0×10^7 y			
95	Americium	241Am	241.0567	*	433 y	710(n,γ) 3(n,f)		
		243Am	243.061373	*	7.37×10^3 y			
96	Curium	242Cm	242.0588	*	163 d	25(n,γ)		
		242Cm	242.0588	*	163 d	25(n,γ) <5(n,f)		
		243Cm	243.06138	*	29.1 y			
		244Cm	244.0628	*	18.1 y	13(n,γ) 1(n,f)		
		247Cm	247.070347	*	1.56×10^7 y			
97	Berkelium	247Bk	247.070299	*				
98	Californium	251Cf	251.079580	*	9.0×10^2 y			
		252Cf	252.08	*	2.65 y (α)	20(n,γ)		
99	Einsteinium	252Es	252.082972	*	1.29 y			
100	Fermium	257Fm	257.095099	*	100.5 d			
101	Mendelevium	258Md	258.098425	*	51.5 d (α)			
102	Nobelium	259No	259.101024	*	58 min			
103	Lawrencium	262Lr	262.109692	*	3.6 h			
104	Rutherfordium	260Rf	260.1064	*	21 ms (sf)			
105	Dubnium	262Db	262.11437	*	34 s			
106	Seaborgium	266Sg	266.12238	*	21 s			
107	Bohrium	264Bh	264.12496	*	0.44 s			
108	Hassium	269Hs	269.1341	*	13 s			
109	Meitnerium	268Mt	268.1388	*	0.05 s			

Appendix II
Mass Attenuation Data, Decay Characteristics of Radioisotopes, and Fast Fission Cross Section for Selected Isotopes

A THE MASS ATTENUATION COEFFICIENT (μ/ρ) FOR SEVERAL MATERIALS, IN cm^2/g, WHERE ρ IS THE NOMINAL DENSITY: DATA OBTAINED FROM *REACTOR PHYSICS CONSTANT*, ANL-5800, 2ND ED., 1963

Material	ρ (g/cm³)	Gamma-Ray Energy (MeV)							
		0.1	0.2	0.5	1.0	3	5	8	10
H	Gas	0.295	0.243	0.173	0.126	0.0691	0.0502	0.0371	0.0321
Be	1.85	0.132	0.109	0.0773	0.0565	0.0313	0.0234	0.0181	0.0161
C	1.60	0.149	0.122	0.0870	0.0636	0.0356	0.0270	0.0213	0.0194
N	Gas	0.150	0.123	0.0869	0.0636	0.0357	0.0273	0.0218	0.0200
O	Gas	0.151	0.123	0.0870	0.0636	0.0359	0.0276	0.0224	0.0206
Na	0.97	0.151	0.118	0.0833	0.0608	0.0348	0.0274	0.0229	0.0215
Mg	1.74	0.160	0.122	0.0860	0.0627	0.0360	0.0286	0.0242	0.0228
Al	2.699	0.161	0.120	0.0840	0.0614	0.0353	0.0282	0.0241	0.0229
Si	2.33	0.172	0.125	0.0869	0.0635	0.0367	0.0296	0.0254	0.0243
K	0.86	0.215	0.127	0.0852	0.0618	0.0365	0.0305	0.0274	0.0267
Ca	1.55	0.238	0.132	0.0876	0.0634	0.0376	0.0316	0.0285	0.0280
Fe	7.87	0.344	0.138	0.0828	0.0595	0.0361	0.0313	0.0295	0.0294
Cu	8.96	0.427	0.147	0.0820	0.0585	0.0357	0.0316	0.0303	0.0305
Pt	21.45	4.75	0.795	0.135	0.0659	0.0414	0.0418	0.0448	0.0477
Pb	11.34	5.29	0.896	0.145	0.0684	0.0421	0.0426	0.0459	0.0489
U	19.1	10.60	1.17	0.176	0.0757	0.0445	0.0446	0.0479	0.0511
Air	0.001293	0.151	0.123	0.0868	0.0636	0.0357	0.0274	0.0220	0.0202
NaI	3.67	1.57	0.305	0.0901	0.0577	0.0367	0.0347	0.0354	0.0366
H_2O	1	0.167	0.136	0.0966	0.0706	0.0396	0.0301	0.0240	0.0219
Concrete	2.25–2.4	0.169	0.124	0.0870	0.0635	0.0363	0.0287	0.0243	0.0229
Tissue	~1	0.163	0.132	0.0936	0.0683	0.0384	0.0292	0.0233	0.0212

B THE MASS ABSORPTION COEFFICIENT (μ_A/ρ) FOR SEVERAL MATERIALS, IN cm²/g, WHERE ρ IS THE NOMINAL DENSITY: DATA OBTAINED FROM *REACTOR PHYSICS CONSTANT*, ANL-5800, 2ND ED., 1963

Material	ρ (g/cm³)	Gamma-Ray Energy (MeV)							
		0.1	0.2	0.5	1.0	3	5	8	10
H	Gas	0.0411	0.0531	0.0591	0.0557	0.0401	0.0318	0.0252	0.0255
Be	1.85	0.0183	0.0237	0.0264	0.0248	0.0283	0.0151	0.0127	0.0118
C	1.60	0.0215	0.0267	0.0297	0.0280	0.0209	0.0177	0.0153	0.0145
N	Gas	0.0224	0.0267	0.097	0.0280	0.0211	0.0180	0.0158	0.0151
O	Gas	0.0233	0.0271	0.0297	0.0280	0.0212	0.0183	0.0163	0.0157
Na	0.97	0.0289	0.0266	0.0284	0.0268	0.0207	0.0185	0.0171	0.0168
Mg	1.74	0.0335	0.0278	0.0293	0.0276	0.0215	0.0194	0.0182	0.0180
Al	2.699	0.0373	0.0275	0.0286	0.0270	0.0212	0.0192	0.0183	0.0182
Si	2.33	0.0435	0.0286	0.0290	0.0274	0.0217	0.0198	0.0190	0.0189
K	0.86	0.0909	0.0340	0.0295	0.0272	0.0222	0.0214	0.0215	0.0219
Ca	1.55	0.111	0.0367	0.0304	0.0279	0.0230	0.0222	0.0225	0.0231
Fe	7.87	0.225	0.0489	0.0294	0.0261	0.0224	0.0227	0.0239	0.0250
Cu	8.96	0.310	0.0594	0.0296	0.0261	0.0223	0.0231	0.0248	0.0261
Pt	21.45	4.646	0.719	0.0892	0.0375	0.0296	0.0343	0.0400	0.0438
Pb	11.34	5.193	0.821	0.0994	0.0402	0.0305	0.0352	0.0412	0.0450
U	19.1	9.63	1.096	0.132	0.0482	0.0332	0.0374	0.0443	0.0474
Air	0.001293	0.0233	0.0268	0.0297	0.0280	0.0211	0.0181	0.0160	0.0153
NaI	3.67	1.466	0.224	0.0410	0.0273	0.0241	0.0268	0.0303	0.0325
H$_2$O	1	0.0253	0.030	0.0330	0.0311	0.0233	0.0198	0.0173	0.0164
Concrete	2.25–2.4	0.0416	0.0289	0.0296	0.0278	0.0216	0.0194	0.0180	0.0177
Tissue	~1	0.0271	0.0293	0.0320	0.0300	0.0220	0.0192	0.0168	0.0160

C DECAY CHARACTERISTICS OF SELECTED RADIOISOTOPES: IT SHOULD BE NOTED THAT BETA ENERGIES ARE THE MAXIMUM ENERGIES AND S.F. STANDS FOR SPONTANEOUS FISSION

Isotope	Half-Life	Type of Decay	Most Predominant Energy(ies) in MeV
H-3	12.6 y	β^-	0.18
Li-8	0.845 s	β^-	13
		2α	3.2 (total)
Be-8	$<1.4 \times 10^{-16}$ s	2α	0.047 (each)
C-14	5715 y	β^-	0.155
N-13	10 min	β^+	1.24
Na-22	2.6 y	β^+, γ	0.542, 1.28
Na-24	15 h	$\beta^-, \gamma_1-\gamma_2$	1.39, 1.37–2.75
Al-28	2.27 min	β^-, γ	2.86, 1.78
P-32	14.3 d	β^-	1.707
K-40	1.2×10^9 y	β^-	1.33 (89%)
		EC, γ	1.46 (11%)
Ca-47	4.8 d	β^-, γ	0.66, 1.3
V-48	16 d	EC, $\beta^+, \gamma_1, \gamma_2$	0.69, 0.986, 1.314
Fe-59	45 d	β^-, γ	0.460, 1.10 (54%)
			0.270, 1.29 (46%)
Co-60	5.24 y	$\beta^-, \gamma_1, \gamma_2$	0.302, 1.33, 1.17
Cu-64	12.9 h	EC, γ	1.34 (42%)
		β^-	0.571 (39%)
		β^+	0.657 (19%)
Cu-66	5.1 min	β^-	2.63 (91%)
		β^-, γ	1.5, 1.04 (9%)
Zn-65	245 d	EC	(55%)
		EC, γ	1.12 (45%)
Kr-85	10.3 y	β^-	0.695 (98.5%)
		β^-, γ	0.15, 0.54 (0.65%)
Sr-90	27.7 y	β^-	0.545
Y-90	64.2 h	β^-	2.26
Ag-108	2.3 min	β^-	1.77 (97.15%)
		EC, γ	0.45 (1.5%)
Ag-110	24 s	β^-	2.82 (40%)
		β^-, γ	2.24, 0.66 (60%)
I-131	8.08 d	β^-, γ	0.608, 0.364 (87.2%)
		β^-, γ	0.335, 0.638 (9.3%)
I-137	22 s	β^-	(94%)
		β^-, n	0.56 (6%)

(Continued)

Isotope	Half-Life	Type of Decay	Most Predominant Energy(ies) in MeV
Xe-135	9.23 h	β^-, γ	0.51, 0.250 (97%)
Xe-137	3.9 min	β^-	3.5
Cs-137	30 y	β^-	0.51 (92%)
		β^-	1.17 (8%)
Ba-133	7.2 y	EC, γ	0.320, 0.081
Ba-140	12.8 d	β^-, γ	1.021, 0.16–0.03 (60%)
		β^-, γ	0.48, 0.54–0.03 (30%)
La-140	40 h	β^-, γ_1–γ_2–γ_3	1.32, 0.33–0.49–1.60 (70%)
		β^-, γ_1–γ_2	1.67, 0.49–1.60 (20%)
Pm-147	2.6 y	β^-	0.223
Ir-192	74.4 d	EC, β^-, γ	0.66, 11 at different energies
Hg-197	65 h	EC, γ	0.077
Pb-214	26.8 min	β^-, γ	0.67, 0.295–0.352
Bi-213	47 min	β^-	1.39 (98%)
		β^-, γ	0.959, 0.434 (2%)
Po-208	2.93 y	α	5.108
Po-210	138 d	α	5.30
Po-213	0.0000042 s	α	8.34
Ra-226	1622 y	α	4.78 (94%)
		α, γ	4.77, 0.186 (5.7%)
Th-231	25.6 h	β^-, γ	0.308, 0.084 (44%)
		β^-, γ	0.094, 0.058, 0.026 (45%)
U-233	1.62×10^5 y	α	4.823
U-234	2.48×10^5 y	α	4.76 (73%)
	s.f. 2×10^{16} y	α, γ	4.72, 0.05 (27%)
U-235	7.1×10^8 y	α	4.40 (83%)
	s.f. 1.9×10^{17} y		
U-238	4.51×10^9 y	α	4.195 (77%)
	s.f. 8×10^{18} y	α, γ	4.18, 0.048 (23%)
U-239	23.5 min	β^-, γ	1.2, 0.73
Np-238	2.1 d	β^-, γ	1.272, 0.044 (47%)
		β^-, γ	0.258, 1.03 (53%)
Pu-238	89.6 y	α	5.49 (72%)
	s.f. 3.8×10^{10} y	α, γ	5.45, 0.044 (28%)
Am-241	458 y	α, γ	5.477, 0.060
Cm-242	35 y	α	6.11 (73.7%)
		α, γ	6.07, 0.044 (26.3%)
Cm-244	17.9 y	α	5.801 (76.7%)
	s.f. 1.4×10^7 y	α, γ	5.759, 0.043 (23.3%)

D FAST FISSION SPECTRUM CROSS SECTIONS OF SELECTED ISOTOPES: DATA SETS ARE OBTAINED FROM R. J. PERRY AND C. J. DEAN, THE WIMS9 NUCLEAR DATA LIBRARY, WINFRITH TECHNOLOGY CENTER REPORT, ANSWERS/WIMS/TR.24, SEPT. 2004

Isotope	Fast Fission Spectrum Cross Section (barns)		
	σ_f	σ_a	σ_s
H-1	0	0.0000392	3.93
B-10	0	0.491	2.12
C-12	0	0.00123	2.36
O-16	0	0.0120	2.76
Na-23	0	0.000234	3.13
Fe-56	0	0.00922	3.20
Zr-91	0	0.00335	5.89
Xe-135	0	0.000743	-
Sm-149	0	0.234	-
Gd-157	0	0.201	6.51
Th-232	0.0713	0.155	7.08
U-233	1.84	1.89	5.37
U-235	1.22	1.29	6.33
U-238	0.304	0.361	7.42
Pu-239	1.81	1.86	7.42
Pu-240	1.36	1.42	6.38
Pu-241	1.62	1.83	6.24
Pu-242	1.14	1.22	6.62

Answers to Selected Problems with Detailed

CHAPTER 1

A. True or False
1. False. No electron
2. True.
3. True.
4. True
5. False. Same mass number.
6. True.
7. False. It should be 1.00×10^3 m
8. True
9. False. It should be 1.8×10^{14}. How?

$$\rho = \frac{m}{V} = \frac{m}{\frac{4}{3}\pi r^3} = \frac{(12 \text{ amu})(1.66 \times 10^{-27} \text{ kg/amu})}{\frac{4}{3}\pi(3 \times 10^{-15} \text{ m})^3} = 1.8 \times 10^{17} \text{ kg/m}^3$$

$$\frac{\rho}{\rho_{water}} = \frac{1.8 \times 10^{17}}{1000} = 1.8 \times 10^{14}$$

10. False. It is the number of atoms in 12 g of C-12.

B. Problems
1. Both have units of 1/second. But Hertz describes frequency and Curie describes the activity of radiation source.
3. 1.398×10^{-24} kg/molecule
5. (a) 238.0186 g/mol   (b) 270.0186 g UO_2/mol   (c) 1.334×10^{18} atoms U-234/cm^3
7. r = 8610.15 cm or 86.1 m

CHAPTER 2

A. True or False
1. True.
2. True.
3. True.
4. False. The de Broglie wavelength should be = 6.63E-34 Js/(0.01 kg × 50 m/s) = 2.65 E-34 m.
5. True
6. False. It is the description of the Compton Scattering effect.

7. True. That is the definition.
8. False. E = hν = (hc)/λ = (1240 eV · nm)/526 nm = 2.36 eV
9. False. Three should be two.
10. True.

B. Problems
1. 1.125 MeV
3. 0.0124 nm
5. 3.38 eV
7. Everything propagates like a wave, exhibiting diffraction and interference, and exchanges energy in discrete lumps like a particle.
9. (a) 1.24 nm   (b) 0.006911 nm   (c) 0.168 MeV

CHAPTER 3

A. True or False
1. True
2. True. He discovered using a magnetic field and photographic plate.
3. True. There is a discovery of alpha particles that should not be coming from the theory of positive charge distribution inside the nucleus.
4. False. It should lie closely to this N = Z line.
5. True. Change of energy = −3.4 − (−13.6) = 10.2 eV.
6. True.
7. False. This rarely happens. More common would be an emission of a small nucleus containing two neutrons and two protons.
8. False. They need an excess of neutrons.
9. True.
10. True.

B. Problems
1. (a) Stable as the number of protons is below 83 and both protons and neutrons are even; (b) Radioactive—although neutrons and protons are less than 83, they are odd and it lies farther from the center of the band of stability; (c) Radioactive as the number of protons is above 83 and both protons and neutrons are odd.
3. 10.2 eV
5. 122 nm for n = 2 to n = 1, 102 nm for n = 3 to n = 1, and 97 for n = 4 to n = 1
7. (a) 1.022 MeV (Note: it is 2 × 0.511 MeV); (b) 1.21×10^{-10} cm; (c) 2.59 MeV
9. Choices (b) and (c) are correct.

CHAPTER 4

A. True or False
1. True
2. True
3. False. It is about 938.3 MeV

4. False. They are not. The rest mass of the proton is roughly 1.007276 u and that of hydrogen is roughly 1.007825 u.
5. False. The sum is greater than the measured mass of a deuterium atom.
6. True
7. True
8. True
9. True
10. False. By calculation, BE of Fe-58 is roughly 509.94 MeV and BE of Ni-62 is roughly 545.26 MeV.

B. Problems

1. 13 eV
3. (a) 11.20 MeV (b) 10.18 MeV (Note that to account for the Mn having one fewer electron than Fe, we replace the mass of the proton with $M(^1H_1)$).
   (c) 26.90 MeV
5. (a) ^{236}U (b) ^1n (c) ^{12}C
6. 2.7945 MeV

CHAPTER 5

A. True or False:

1. True
2. True
3. True
4. False. Based on the description of the problem, you should see oxygen-17. This is the observation reported by Rutherford.
$$^4_2He + ^{14}_7N \rightarrow ^{17}_8O + ^1_1H \quad \text{or} \quad ^{14}_7N(\alpha, p)^{17}_8O$$

5. True. This is discovered by Chadwick in 1952.
6. False. Elastic scatter or KE is not conserved.
7. True.
   This is $^{30}P \rightarrow ^{30}Si +?$
   So, the sum of protons: LHS 15   RHS $14 + x$...thus $x = 1$
   So, the sum of mass: LHS 30   RHS $30 + y$ ... thus $y = 0$
   Hence the particle must be a positron.
8. False. Energy must be conserved.
9. False. Mass is converted to energy.
10. True.

B. Problems

1. (a) Helium-3   (b) Beryllium-9   (c) Carbon-14
3. BE of He-4 is 28.294 MeV. BE of deuterium is 2.224 MeV. BE of He-3 is 7.717 MeV. The Q-value is −18.353 MeV. So, the BE of the proton is 0 MeV. The Overall reaction is endothermic.
5. $^{206}Pb_{82}$
7. 3.27 MeV and this is exothermic.
9. 4.917 MeV

CHAPTER 6

A. True or False:

1. True
2. False. The description is for internal conversion. Electron capture happens when an orbital electron is absorbed by the nucleus and converts a nuclear proton into a neutron and neutrino. This will leave the nucleus in an excited state.
3. False. An anti-neutrino is emitted during the negatron decay.
4. True
5. True
6. True
7. False. It is a statistical process.
8. False. If the isotope is initially 100 atoms, after 4 days, it will be 50 atoms and 8 days, it will be 25 atoms, 12 days → 12.5 atoms, and 16 days → 6.25 atoms. It will not yet reach zero.
9. False. The first part is correct. But the second part is incorrect as the activity will be the same for both parents and daughters.
10. True. 1 Curie = 3.7E10 disintegration/sec and 1 Bq = 1 disintegration/second.

B. Problems

1. (a) $^{22}Na_{11} \rightarrow {}^{22}Ne_{10} + {}^{0}e_{+1} + \nu_e$ (b) In the decay reaction $^{38}Cl_{17} \rightarrow {}^{38}Ar_{18} + {}^{0}e_{-1} + \nu_e^*$
   (c) $^{224}Ra \rightarrow {}^{220}Rn + \alpha$ (d) $^{41}Ar_{18} \rightarrow {}^{41}K^*_{19} + {}^{0}e_{-1} + \nu$
3. 5.686 MeV for the 94% probability and 5.449 for the 5.5 probability
5. (a) 5.90×10^{-7} s^{-1} (b) 1.17×10^6 seconds   (c) 1.69×10^6 seconds
7. (a) 3.13×10^{10} Bq   (b) 3.13×10^{10} Bq   (c) 1.55×10^{-6} g
   (d) 2.90×10^9 dis/s = 2.90×10^9 Bq
9. 10.8 hours

CHAPTER 7

A True or False

1. False. FF are radioactive.
2. False. Based on the fission yields, the range of a given mass number has the lowest fission yield.
3. True.
4. False. The energy carried away by the neutrinos is lost and mainly the total available as heat comes from FF and neutron KE and primary radiations.
5. True.
6. False. The two are the kinematic and Coulomb barrier types.
7. True. With the formula for lethargy = 2/(A + 2/3), we definitely can see that the lethargy of helium should be way larger than uranium.
8. True. Since gamma is given off in radiative capture, shielding can be cumbersome and quite hazardous.
9. False. It is way faster than that, roughly 10–14 seconds.
10. False. It is <1% for thermal fission. The generation description is correct.

Shultis & Faw Problem 6.4   **Minimum**

n	Li6	1st prod	2nd prod	Delta	Q	Eth	KE of products
1.008665	6.015122	7.016004	0	0.007783	7.250	0.000	7.250
1.008665	6.015122	6.015122	1.008665	0	0.000	0.000	0.000
1.008665	6.015122	6.018888	1.007825	−0.00293	−2.726	3.180	0.454
1.008665	6.015122	5.01222	2.014102	−0.00254	−2.361	2.755	0.394
1.008665	6.015122	3.016049	4.002603	0.005135	4.783	0.000	4.783

B. Problems

1. (a) 7.250 MeV   (b) 3.180 MeV
   (c)The minimum kinetic energy imparted to the products is the sum of the Q value and the applicable kinematic threshold energy should the reaction be endothermic.
3. (a) $^6Li_3 + {}^1n_0 \rightarrow {}^3H_1 + {}^4He_2$ and $\quad {}^{16}O_8 + {}^3H_1 \rightarrow {}^{18}F_9 + {}^1n_0$
   (b) 4.78 MeV and 1.27 MeV   (c) 2.42 MeV   (d) 2.73 MeV. Since this is greater than the Coulombic threshold energy of 2.42 MeV for the second reaction, the answer is **Yes**, the reaction is feasible with thermal neutrons.
5. 2.21×10^9 n/s
7. 0.476 MeV, 0.444 MeV, 0.416 MeV, and 1.86 MeV
9. (a) 183.6 MeV (b) 6.80×10^{18} fissions/s (c) 3.12×10^{-3} g/s

CHAPTER 8

A. True or False

1. True. Particles are the neutron and a nucleus is the target area.
2. True.
3. False. Absorption = capture + fission. So, the fission should be 20 b.
4. True.
5. False. It is flux. Fluence needs to include the time duration it takes as well.
6. True.
7. True.
8. False. It is followed the inverse square law.
9. True.
10. True. Gamma uses the attenuation coefficient. Neutrons on another hand will use the macroscopic cross-section.

B. Problems

1. (a) 0.20 cm^{-1} (b) 5 cm
3. 0.242 cm^{-1}
5. (a) $dN_A/dt = R_A - \lambda_A N_A$
   $dN_B/dt = R_B + \lambda_A N_A - \lambda_B N_B$
   $dN_C/dt = \lambda_B N_B - \lambda_C N_C$

$dN_D/dt = f\lambda_C N_C - \lambda_D N_D$

$dN_E/dt = R_E + \lambda_D N_D$

$dN_F/dt = (1 - f)\lambda_C N_C$

(b) $N_A(t) = (R/\lambda)(1 - e^{-\lambda t}) + N_{0A}e^{-\lambda t}$

Essentially, the solution can be broken into two parts. The first part, (R/λ) $(1 - e^{-\lambda t})$, tracks the population of those atoms which did not exist at time zero but are being created at a constant rate R. The solution is identical to that given near the end of the Chapter 5 notes for the saturation activity of an isotope which is being created at a constant rate R, but whose initial number at time zero is zero, that is, $N(t = 0) = 0$.

The second part, $N_{0A}e^{-\lambda t}$, tracks the history of those atoms that initially existed at time zero, that is, N_0. This is simply the solution to the case of a radioactive sample whose atoms are not being replenished.

7. (a) 6.0×10^9 cm^{-2}-s^{-1} (b) 4.0×10^9 cm^{-2}-s^{-1}

9. (a) n = 66.7 cm^{-3} (b) n = 9.09×10^6 cm^{-3}

11. (a) 9.81 cm   (b) 1.48 cm   (c) 0.898 cm

CHAPTER 9

A. True or False

1. False. A good moderator should have a high MR and a large MSDP.
2. False. This is inversely proportional to the square root of temperature. Thus, it should decrease.
3. True.
4. False. This is not the case. It is independent of temperature. The ratio is 1.128.
5. False. It is independent of the geometry and only relies on the material properties.
6. True.
7. True. N-235/N-238 = 0.90/0.10 = 9
8. False. The thermal utilization factor is the ratio of the number of neutrons absorbed in the fuel to the total number of absorptions in fuel, moderator, cladding, etc. Therefore, adding more moderator quantity will decrease this factor.
9. True.
10. False. Lethargy is ~ 2/(A + 2/3). Therefore, the amount of graphite does not affect its value.

B. Problems

1. (a) 0.0253 eV   (b) 0.0259 eV (close enough)
3. 0.186 b
5. 0.020
7. 1.61 cm
9. (a) −45°C   (b) decrease
11. 4254
13. 205.88 or 206 cm.

CHAPTER 10

A. True or False

1. False. There has been tested with enriched uranium.
2. False. Oxide fuel is being used.
3. False. PWR control rod – downward. BWR control rod – upward
4. True. TCE = PE/PT
5. True. About 40% in GCR and only 34% in LWR.
6. False. Not giving off radiation. Giving off neutrons.
7. True. Also, larger lethargy can be used as a coolant.
8. False. Water has a low boiling point. Therefore, the pressure vessel is necessary and this becomes cumbersome in size.
9. True.
10. False. MSBR is one of the six proposed designs in the Gen-IV group.

B. Problems

1. (a) 3030.30 MWt and 1030.30 MWe   (b) 31897.9 kg/s
3. (a) The major change will be in the moderator thermal absorption cross-section. We see that the absorption cross section in heavy water is much smaller than for water. So, the thermal utilization factor would decrease.
(b) Because of its very small thermal abortion cross section, heavy water is considered to be the best moderator; reactors can be built using natural uranium if heavy water is the moderator, but not with ordinary water. Thus, the net effect of replacing heavy water with water would be to decrease the value of $k\infty$.
5. Choice d
7. Choice b
9. Choice b

CHAPTER 11

A. True or False

1. True. R is the quantity of X- or gamma radiation such that it associated corpuscular emission per 0.001293 g of air produces, in air, ions carrying 1 esu of the quantity of electricity of either sign. It has no time involved.
2. True. This is the correct conversion
3. False. It should be 1 Gy = 1 J/kg = 100 rad.
4. False. Rem is correct. But Sievert is used instead of Gray for biological systems.
5. True.
6. False. The quality factor for alpha is about 20. And the QF for beta is ~1.
7. True. It increases as elevation increases. Teton is around 13,777 ft compared to the sea level in Richmond, VA.
8. True. The reproduction cell is even more sensitive than the fetus.
9. True.
10. False. K ~> D. This is because the energy absorption coefficient used in the absorbed dose calculation is smaller than the energy transfer coefficient used in the kerma calculation.

B. Problems

1. QF of the thermal neutron is 5. So, D = 1.67 mrad, H = 8.35 mrem and f = 8.35 mrem/5000 mrem = 1.67×10^{-3}

3. You would select **boron** since alpha-particles ($^4He_2$) have a limited mean free path even in the air so the radiation exposure would be limited compared to that of the high-energy gamma rays from the cadmium reaction.

5. (a) 4.2 mSv/hour   (b) 2.1 mSv or 210 mrem

7. Low to high order: Muscle, bone, thyroid, skin, intestines, and fetus

9. 0.001 Ci or 1 mCi

Index

toxicity 252
transport mean free path 208
The uncertainty principle 30

uncollided flux 168, 173–175

virtual state 114
volume fractions 155

waste management 239
wave–particle duality 30
weight (mass) fractions 155
work function 21, 158

X-ray 24–25, 245

Yellowcake 239

Printed in the United States
by Baker & Taylor Publisher Services